Lecture Notes on Data Engineering and Communications Technologies

Volume 27

Series editor

Fatos Xhafa, Technical University of Catalonia, Barcelona, Spain
e-mail: fatos@cs.upc.edu

The aim of the book series is to present cutting edge engineering approaches to data technologies and communications. It will publish latest advances on the engineering task of building and deploying distributed, scalable and reliable data infrastructures and communication systems.

The series will have a prominent applied focus on data technologies and communications with aim to promote the bridging from fundamental research on data science and networking to data engineering and communications that lead to industry products, business knowledge and standardisation.

More information about this series at http://www.springer.com/series/15362

Isaac Woungang • Sanjay Kumar Dhurandher
Editors

2nd International Conference on Wireless Intelligent and Distributed Environment for Communication

WIDECOM 2019

Springer

Editors
Isaac Woungang
Department of Computer Science
Ryerson University
Toronto, ON, Canada

Sanjay Kumar Dhurandher
CAITFS, Department of Information
Technology
Netaji Subhas University of Technology
New Delhi, India

ISSN 2367-4512   ISSN 2367-4520   (electronic)
Lecture Notes on Data Engineering and Communications Technologies
ISBN 978-3-030-11436-7   ISBN 978-3-030-11437-4   (eBook)
https://doi.org/10.1007/978-3-030-11437-4

Library of Congress Control Number: 2019934719

© Springer Nature Switzerland AG 2019
This work is subject to copyright. All rights are reserved by the Publisher, whether the whole or part of
the material is concerned, specifically the rights of translation, reprinting, reuse of illustrations, recitation,
broadcasting, reproduction on microfilms or in any other physical way, and transmission or information
storage and retrieval, electronic adaptation, computer software, or by similar or dissimilar methodology
now known or hereafter developed.
The use of general descriptive names, registered names, trademarks, service marks, etc. in this publication
does not imply, even in the absence of a specific statement, that such names are exempt from the relevant
protective laws and regulations and therefore free for general use.
The publisher, the authors, and the editors are safe to assume that the advice and information in this book
are believed to be true and accurate at the date of publication. Neither the publisher nor the authors or
the editors give a warranty, express or implied, with respect to the material contained herein or for any
errors or omissions that may have been made. The publisher remains neutral with regard to jurisdictional
claims in published maps and institutional affiliations.

This Springer imprint is published by the registered company Springer Nature Switzerland AG.
The registered company address is: Gewerbestrasse 11, 6330 Cham, Switzerland

Welcome Message from WIDECOM 2019 General Chair

Welcome to the 2nd International Conference on Wireless, Intelligent, and Distributed Environment for Communication (WIDECOM 2019).

The last decade has witnessed tremendous advances in computing and networking technologies, with the appearance of new paradigms such as Internet of Things (IoT) and cloud computing, which have led to advances in wireless and intelligent systems for communications. Undoubtedly, these technological advances help improve many facets of human lives, for instance, through better healthcare delivery, faster and more reliable communications, significant gains in productivity, and so on. At the same time, the associated increasing demand for a flexible and cheap infrastructure for collecting and monitoring real-world data nearly everywhere, coupled with the aforementioned integration of wireless mobile systems and network computing, raises new challenges with respect to the dependability of integrated applications and the intelligence-driven security threats against the platforms supporting these applications. The WIDECOM conference is a conference series that provides a venue for researchers and practitioners to present, learn, and discuss recent advances in new dependability paradigms, design, and performance of dependable network computing and mobile systems, as well as issues related to the security of these systems.

Every year, WIDECOM receives dozens of submissions from around the world. Building on the success from last year, WIDECOM 2019 presents an exciting technical program that is the work of many volunteers. The program consists of a combination of technical papers, keynotes, and tutorials. The technical papers are peer-reviewed by program committee members who are all experts and researchers, through a blind process.

We received a total of 38 papers this year and accepted 17 papers for inclusion in the proceedings and presentation at the conference, which corresponds to an acceptance rate of about 45%. Papers were reviewed by two PC members, in a single round of review.

WIDECOM 2019 is also privileged to have select guest speakers to provide stimulating presentations on topics of wide interest. This year's distinguished speakers are

- Dr. Elena Pagani, Associate Professor, Computer Science Department, University of Milan, Italy, and Associate Researcher at the Institute for Informatics and Telematics of the National Research Council in Pisa, Italy
- Dr. Francesco Bruschi, Assistant Professor, Dipartimento di Elettronica Informazione e Bio-ingegneria, Politecnico di Milano, Milan, Italy

We would like to thank all of the volunteers for their contributions to WIDECOM 2019. Our thanks go to the authors, and our sincere gratitude goes to the program committee, who gave much extra time to carefully review the submissions.

We are pleased to announce selected papers will be invited to submit extended versions for publication in the *International Journal of Space-Based and Situated Computing* (IJSSC), Inderscience; *International Journal of Grid and Utility Computing* (IJGUC), Inderscience; and *Internet of Things: Engineering Cyber Physical Human Systems*, Elsevier.

We would also like to thank the organizing committee, in particular, the local organizing team Michela Ceria, Luca Casati, and Alessandro De Piccoli, all from the University of Milan, for their support and hard work in making this event a success. Our thanks go to our sponsors:

- Computer Science Department, University of Milan, Italy, for hosting WIDECOM 2019
- Springer, for publishing the conference proceedings

Finally, we thank all the attendees and the larger WIDECOM 2019 community for their continuing support, by submitting papers and by volunteering their time and talent in other ways.

We hope you will find the papers presented interesting and enjoy the conference.

Milan, Italy

Andrea Visconti
WIDECOM 2019 Conference Chair

Welcome Message from the WIDECOM 2019 Program Cochairs

Welcome to the 2nd International Conference on Wireless, Intelligent, and Distributed Environment for Communication (WIDECOM 2019), which was held from February 11 to February 13, 2019, at University of Milan, Italy.

WIDECOM 2019 provides a forum for researchers and practitioners from industry and government to present, learn, and discuss recent advances in new dependability paradigms, design, and performance of dependable network computing and mobile systems, as well as issues related to the security of these systems.

The papers selected for publication in the proceedings of WIDECOM 2019 span many research issues related to the aforementioned research areas, covering aspects such as algorithms, architectures, protocols dealing with network computing, ubiquitous and cloud systems and Internet of Things systems, integration of wireless mobile systems and network computing, and security. We hope the participants to this conference will benefit from this coverage of a wide range of current hop-spot-related topics.

In this edition, 38 papers were submitted and peer-reviewed by the program committee members and external reviewers who are experts in the topical areas covered by the papers. The program committee accepted 17 papers (about 45% acceptance ratio). The conference program also includes two distinguished keynote speeches and three tutorials.

Our thanks go to the many volunteers who have contributed to the organization of WIDECOM 2019. We would like to thank all authors for submitting their papers. We would also like to thank the program committee members for thoroughly reviewing the submission and making valuable recommendations. We would like to thank WIDECOM 2019 local arrangement team for the excellent organization of the conference and for their effective coordination creating the recipe for a very successful conference.

We hope you will enjoy the conference and have a great time in Milan, Italy.

Toronto, ON, Canada	Isaac Woungang
New Delhi, India	Sanjay Kumar Dhurandher
	WIDECOM 2019 Program Committee Cochairs

WIDECOM 2019 Organizing Committee

General Chair:
- Andrea Visconti, University of Milan, Italy

Local Organizing Chairs:
- Michela Ceria (Chair), University of Milan, Italy
- Luca Casati, University of Milan, Italy
- Alessandro De Piccoli, University of Milan, Italy

Workshop and Publicity Cochairs:
- Nitin Gupta, University of Delhi, India
- Udai Pratap Rao, S. V. National Institute of Technology, India

Tutorial Chair:
- Nitin Gupta, University of Delhi, India

TPC Cochairs:
- I. Woungang, Ryerson University, Canada
- S. K. Dhurandher, Netaji Subhas University of Technology, India

Technical Program Committee:
- Michela Ceria, University of Milan, Italy
- Nadir Murru, University of Torino, Italy
- Federico Pintore, University of Oxford. UK
- Changyu Dong, Newcastle University, UK
- Joel Rodrigues, University of Beira Interior, Portugal
- Vinesh Kumar, University of Delhi, India
- Glaucio Carvalho, Ryerson University
- Chii Chang, University of Tartu, Estonia
- Petros Nicopolitidis, Aristotle University of Thessaloniki, Greece
- Wei Lu, Keene State College, USA
- Zeadally Sherali, University of Kentucky, USA
- Luca Caviglione, CNIT, Italy
- Hamed Aly, Acadia University, Canada

- Rohit Ranchal, IBM Watson Health Cloud, USA
- Ramilo Liscano, University of Ontario Institute of Technology, Canada
- Tom Walingo, University of Pretoria, South Africa
- Isaac Woungang, Ryerson University
- Sanjay K. Dhurandher, Netaji Subhas University of Technology

WIDECOM 2019 Keynote Talks

From WSNs to VANETs: Paradigms, Technologies, and Open Research Issues for Challenged Networks

Elena Pagani

E. Pagani
Associate Professor, Computer Science Department, University of Milan, Italy

Associate Researcher at the Institute for Informatics and Telematics of the National Research Council in Pisa, Pisa, Italy

Abstract

Wireless technologies are going to revolutionize countless aspects of daily life, touching issues ranging from smart cities to opportunistic networks, from smart vehicles to Industry 4.0. At the base of all these applications, there is the requirement of implementing communication services among devices possibly with scarce resources and also mobile. The purpose of this talk is to analyze the paradigms for mobile ad hoc networking and the characteristics of these infrastructures, to discuss the state of the art of the technologies and solutions available to implement them, and to highlight the research aspects that are still open.

Making Sense(s) of Smart Contracts in a Connected World

Francesco Bruschi

F. Bruschi
Assistant Professor, Dipartimento di Elettronica Informazione e Bio-ingegneria, Politecnico di Milano, Milan, Italy

Abstract

Smart contracts are digital, executable, self-enforcing descriptions of commitments between parts. They were first envisioned in 1994 by Nick Szabo and recently, blockchain-based platforms such as Ethereum, with their very strong guaranties of untampered, deterministic execution, seem to offer an ideal medium for their deployment. Among the main applications of smart contracts is the automatization of interactions such as bets, collaterals, prediction markets, and insurances. One of the main issues that arise in extending the domain of smart contracts regards provisioning reliable information on the blockchain (the so-called oracle problem). In this talk, we will consider the challenges that emerge when using IoT sensors and devices as information providers for smart contracts.

WIDECOM 2019 Tutorials

Tutorial 1: High-Speed Cryptography

Alessandro De Piccoli

A. De Piccoli
PhD student, Computer Science Department, University of Milan, Milan, Italy

Abstract

In today's communications, cryptography is increasingly crucial to ensure the confidentiality of the exchange of data, whether they are related to commercial transactions or simply information. Exploiting the recent optimization of the product between polynomials, especially those having binary coefficients, it is possible to improve the implementation of cryptography algorithms. In fact, when implementing a cryptographic algorithm, efficient operations have high relevance both in hardware and software. Since a number of operations can be performed via polynomial multiplication, the arithmetic of polynomials over finite fields plays a key role in real-life implementations. Through new techniques, we can save time and memory, allowing fast encryption/decryption operations but also faster cryptanalysis. At the same time, this technique may be used by researchers to prevent attacks such as side-channel and timing attacks. The aim of this tutorial is to provide an introductory guide to the mathematical aspects of high-speed cryptography techniques for computer science researchers.

Tutorial 2: Understanding the Key Pre-distribution Aspect of Linear Wireless Sensor Network

Kaushal Shah

K. Shah
Assistant Professor at CGPIT, Uka Tarsadia University, Surat, Gujarat, India

Abstract

There is a set of applications in wireless sensor networks that forms a particular topology through specific placements of sensor nodes. This set is known as linear infrastructure or one-dimensional network. Applications of such networks are subway tunnel or pipeline monitoring and perimeter surveillance. These applications often demand critical security concerns. The distribution of symmetric keys for such networks is different from those that are planned and are widely studied. By considering requirements for such linear infrastructure in detail, we observe that connectivity is an important issue as capturing a single node disrupts the entire network's services. Therefore, we propose a new measure of connectivity that produces the optimal results with acutely lightweight key pre-distribution schemes (KPS). We also show the theoretical analysis of our proposed scheme and prove how it produces optimal results with lesser storage requirements as compared to other schemes. The performance analysis shows that the proposed KPS requires lesser number of keys per node ($O(1)$ constant storage), as compared to the other existing schemes in the literature, to provide the same level of connectivity.

Tutorial 3: Efficient Cryptographic Algorithms for Securing Passwords

Michela Ceria

M. Ceria
Postdoc Fellow, Computer Science Department, University of Milan, Milan, Italy

Abstract

Nowadays, there are several real-life applications that require authentication and so the use of passwords: e-mails, online banking, mobile phones, and so on. A server retaining our passwords must ensure them the highest possible level of security from attacks (e.g., dictionary attacks and brute force attacks), also taking into consideration that the average user is not able to generate a strong password with a suitable number of entropy bits (the average password entropy was estimated at 40.54 bits) and that more than one user likely employ the same password for the same kind of service. This problem can be addressed by means of cpu- and memory-intensive algorithm that slow attackers down. In this talk, we will give an overview of main password hashing algorithms such as Argon2, Catena, Lyra2, Makwa, and yescrypt, explaining their main features and instances.

Contents

1 Performance Evaluation of G.711 and GSM Codecs on VoIP Applications Using OSPF and RIP Routing Protocols 1
Nadia Aftab, Maurin Hassan, Muhammad Nadeem Ashraf, and Akash Patel

2 Cyclic Redundancy Check Based Data Authentication in Opportunistic Networks 17
Megha Gupta

3 Hybrid Cryptographic Based Approach for Privacy Preservation in Location-Based Services 27
Ajaysinh Rathod and Vivaksha Jariwala

4 Design of Energy-Aware PRoPHET and Spray-and-Wait Routing Protocols for Opportunistic Networks 35
Sibusiso Shabalala, Zelalem Shibeshi, and Khuram Khalid

5 An Asymmetric RSA-Based Security Approach for Opportunistic IoT 47
Nisha Kandhoul and Sanjay Kumar Dhurandher

6 Performance Analysis of A*-Based Hop Selection Technique in Opportunistic Networks Through Movement Mobility Models 61
Pragya Kuchhal and Satbir Jain

7 Data Loss Prevention Using Document Semantic Signature 75
Hanan Alhindi, Issa Traore, and Isaac Woungang

8 Understanding Optimizations and Measuring Performances of PBKDF2 101
Andrea Francesco Iuorio and Andrea Visconti

9 PSARV: Particle Swarm Angular Routing in Vehicular Ad Hoc Networks 115
Mrigali Gupta, Nakul Sabharwal, Priyanka Singla, Jagdeep Singh, and Joel J. P. C. Rodrigues

10 A Reliable Firefly-Based Routing Protocol for Efficient Communication in Vehicular Ad Hoc Networks 129
Nakul Sabharwal, Priyanka Singla, Mrigali Gupta, Jagdeep Singh, and Joel J. P. C. Rodrigues

11 Exploring the Application of Random Sampling in Spectrum Sensing 143
Hayat Semlali, Najib Boumaaz, Asmaa Maali, Abdallah Soulmani, Abdelilah Ghammaz, and Jean-François Diouris

12 White-Box Cryptography: A Time-Security Trade-Off for the SPNbox Family 153
Federico Cioschi, Nicolò Fornari, and Andrea Visconti

13 *CESIS*: Cost-Effective and Self-Regulating Irrigation System 167
Kaushal A. Shah, Meet Patel, Monil Khasakiya, Saad Kazi, and Pinkesh Khalasi

14 Maximum Eigenvalue Based Detection Using Jittered Random Sampling 183
Asmaa Maali, Sara Laafar, Hayat Semlali, Najib Boumaaz, and Abdallah Soulmani

15 Prevention of Flooding Attacks in Mobile Ad Hoc Networks 193
Gurjinder Kaur, V. K. Jain, and Yogesh Chaba

16 Exploiting ST-Based Representation for High Sampling Rate Dynamic Signals 203
Andrea Toma, Tassadaq Nawaz, Lucio Marcenaro, Carlo Regazzoni, and Yue Gao

17 Real-Time Spectrum Occupancy Prediction 219
S. A. Abdelrahman, Omar Khaled, Amr Alaa, Mohamed Ali, Injy Mohy, and Ahmed H. ElDieb

18 SCC-LBS: Secure Criss-Cross Location-Based Service in Logistics 233
Udai Pratap Rao, Gargi Baser, and Ruchika Gupta

Index 257

Chapter 1
Performance Evaluation of G.711 and GSM Codecs on VoIP Applications Using OSPF and RIP Routing Protocols

Nadia Aftab, Maurin Hassan, Muhammad Nadeem Ashraf, and Akash Patel

1.1 Introduction

VoIP is an emerging technology which involves sending voice packets over an IP network rather than via a public switched telephone network (PSTN)—which uses the traditional circuit switching telephony. VoIP requires low packet loss and low delay to ensure voice quality. Therefore, when implementing VoIP solutions, many factors should be taken into account, including the choice of codec, the routing protocol used, the quality of service (QoS), the delay variation, and the packet loss, just to name a few [1]. Many companies have shifted to VoIP solutions or a hybrid model of PSTN and VoIP in order to save their operational cost by utilizing the data bandwidth. There is a dilemma of designing and optimizing networks to address the ongoing requirement of adding VoIP traffic, which motivated the desire to take a closer look at VoIP solutions.

In real world, an IP-based network consists of various types of traffic, including VoIP, HTTP, FTP, and UDP. In this chapter, our focus is only on VoIP traffic. A key factor for VoIP is QoS, which is meant to ensure that voice packets are not dropped or delayed during the transmission, a call, or an event. VoIP is a transfer of voice packets, which is a real-time application that is very sensitive to delay and jitter. For this reason, QoS has become a very important criterion for VoIP quality. In order to upgrade the quality of voice packet transmission, VoIP QoS can be analyzed and configured using a combination of different parameters such as packet loss, end-to-end delay, echo, and jitter.

N. Aftab (✉) · M. Hassan · M. N. Ashraf · A. Patel
Faculty of Engineering and Architectural Science, Ryerson University, Toronto, ON, Canada
e-mail: naftab@ryerson.ca; maurin.hassan@ryerson.ca; m10ashraf@ryerson.ca;
akpatel@ryerson.ca

© Springer Nature Switzerland AG 2019
I. Woungang, S. K. Dhurandher (eds.), *2nd International Conference on Wireless Intelligent and Distributed Environment for Communication*, Lecture Notes on Data Engineering and Communications Technologies 27,
https://doi.org/10.1007/978-3-030-11437-4_1

In this chapter, different QoS parameters have been defined and analyzed in order to carry out our simulations. The performance of two different routing protocols (namely, RIP and OSPF) in IPv4 is studied using two different voice encoding codecs, namely G.711 and GSM, and the Riverbed Modeler version 17.5 [2], considering the jitter, end-to-end delay, packet delay variation, and MOS value as performance metrics. Simulation results exhibit the superiority of G.711 over GSM as preferred VoIP solution in terms of the aforementioned metrics.

The rest of the chapter is organized as follows. In Sect. 1.2, some background on Voice Encoding Codecs and considered IP Routing protocols are presented. In Sect. 1.3, related works are discussed. In Sect. 1.4, the network design, configuration parameters, and considered routing scenarios are described. In Sect. 1.5, the performance evaluation of the proposed design is conducted. Finally, Sect. 1.6 concludes the chapter.

1.2 Background

1.2.1 Voice Encoding Codec

Voice transmission is analog, whereas a communication network is digital. The process or algorithm to convert analog voice signal into digital information is performed by an encoder–decoder device (codec). The codec samples the analog signal according to the algorithm into the digital data to be transmitted over a network or the Internet. A codec generally compresses the data to improve the performance by consuming less bandwidth and without degrading the quality (lossless compression). In this process, the data may also be encrypted to achieve security while transmitting it over the network. Hence, a codec is expected to achieve the following three criteria [3]:

- Encoding and decoding
- Compression and decompression
- Encryption and decryption

Many different voice encoding codecs have been proposed in the literature, including G.722, G.729, and G.726 [3]. In this chapter, we only focus on GSM and G.711. Their features are given in Table 1.1.

1.2.2 Routing Protocols

IP routing is an important component of modern data networks. VoIP is a Layer 3 network protocol that uses Layer 2 point-to-point or link-layer protocols [4]. Selecting an appropriate routing protocol is crucial for VoIP network in order to

Table 1.1 Types of voice encoding codec

GSM	G.711
First digital voice encoding standard used in digital mobile phone systems	Narrow band audio codec primarily used in telephony
Also known as full rate	Also known as pulse code modulation (PCM)—a commonly used waveform codec
Quality of coded speech is poor	Provides toll quality audio Passes audio signals in the range of 300–3400 Hz[a] and samples at a rate of 8000 samples/s
Bit rate of codec: 13 kbits/s[a]	Bit rate of codec: 64 kbits/s[a]
It is a good compromise between computational complexity and quality	Lossless data compression with decrease in audio quality and bandwidth
Codec still widely in use	A required standard for many applications

[a]*kbits/s* Kilobits per second, *Hz* Hertz

ensure efficient route convergence as it influences the jitter and latency [5]. There are a number of IP routing protocols available such as IGRP, EIGRP, RIP, OSPF, and IS-IS, each with its own advantages and disadvantages. Depending on the algorithm used to find the best route between two nodes, most of them can be categorized into two classes: distance vector and link state. In this work, we have considered RIP and OSPF.

The Routing Information Protocol (RIP) is a distance-vector protocol based on Bellman and Ford algorithm [6], which sets a limitation on the number of hops allowed in a path from source to destination to prevent routing loops. It is easier to configure than OSPF or IS-IS, but generally yields a poor convergence. RIP allows routers to update their routing tables at periodic timeframes, most commonly at 30-s intervals. Typically, it broadcasts the entire routing table to all connected neighboring routers with all advertisement and generates high traffic on the network.

On the other hand, the Open Shortest Path First (OSPF) [6] is a link-state routing protocol which is based on the Dijkstra's algorithm. Typically, it computes the shortest path tree from source to destination, and then sends out the network updates to all neighbors. Upon receipt of these updates, each neighbor relays them to each of its own neighbors. Then, each router gathers this information in order to build a map of the entire network so as to enable communication and data transfer among all nodes.

1.3 Related Work

Call quality for VoIP technology is dependent on the design and implementation of the VoIP network that will be used. In [1], Syed and Ambore reported that the codec choice is the first factor in determining the quality of a VoIP call, and a codec with higher bit rate generally yields a better voice quality. They also argued that G.711 is the best available codec for voice encoding. In [5], Che and Cobley showed how VoIP performance can be affected by different routing protocols in the IPv4 technology. They simulated the RIP, OSPF, and EIGRP under three different scenarios using OPNET modeler. To study the route reconvergence, which is an important factor to VoIP user as it impacts both the latency and jitter, they configured a link to fail and autorecover. Their results show that OSPF is more consistent throughout the process, making it a better choice for VoIP solution compared to RIP and EIGRP. In [7], Singh et al. reported that signal quality in a VoIP solution depends on various factors such as networking conditions, coding processes, and speech content. In [8], Hussein and Jamwal studied the RIP, OSPF, and EIGRP protocols in a hybrid Ring and Star topology based on IPv4 technology using Riverbed Modeler Academic Edition 17.5. Their simulation results showed that EIGRP converges faster in large network (5-s) compared to OSPF (7.2-s) and RIP (14.15-s). To evaluate the performance of VoIP network, Ahmed et al. [9] used different metrics such as MOS, packet delay, end-to-end packet delay variation, IP traffic dropped, and Jitter. In [10], Besacier et al. focused on identifying the performance of audio and speech compression; they reported that MPEG transcoding impairs the speech recognition for low bitrates and sustains the performance of speech coders such as GSM and G.711. In [11], Sathu and Shah studied the performance of G.711.1, G.711.2, G.723.1, G.729.2, and G.729.3 codecs in VoIP, and reported G.711.1 as the preferred codecs to be used for both Windows and Linux operating systems. Their simulation results also revealed that Windows 7 generally yields a lesser delay when the G.729.2 and G.729.3 codecs are utilized.

1.4 Network Model

1.4.1 Network Topology

The topology considered in this chapter is a small enterprise network consisting of four CISCO 7200 routers (labeled as R1, R2, R3, and R4), two Nortel PS 6480 Ethernet switches (labeled as Switch 1 and Switch 2), and two IP phones (labeled as IP Phone1 and IP Phone2). All routers and switches are connected to each other via point-to-point (PPP) Digital Signal 3 (DS3) data links. The IP Phones are only supported via 10, 100, or 1000 BASE links; therefore, the highest available link 1000 BASE-X is used to connect them to switches. This topology is depicted in Fig. 1.1. It should be noted that the "Auto-Assign IP Addresses" option is used to assign the IPv4 address for all interfaces.

The considered topology [1] depicts a simple VoIP network exchanging voice packets to communicate between two IP phones, which are themselves connected via Ethernet switches and routers (as per Fig. 1.1).

1 Performance Evaluation of G.711 and GSM Codecs on VoIP Applications... 5

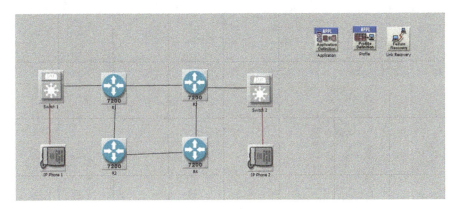

Fig. 1.1 Network topology

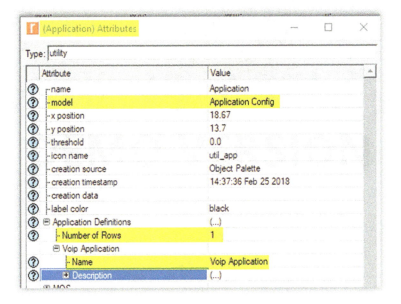

Fig. 1.2 Application configuration

1.4.2 Configuration Parameters

The application definition (Fig. 1.2), profile definition (Fig. 1.3), and failure recovery (Fig. 1.6) object are configured from the object palette to utilize the VoIP application.

Under the application attributes, the G.711 encoder scheme and PCM quality speech for voice are used for scenarios that are running G.711 (as shown in Fig. 1.4). On the other hand, the GSM encoder scheme and GSM quality speech for voice are

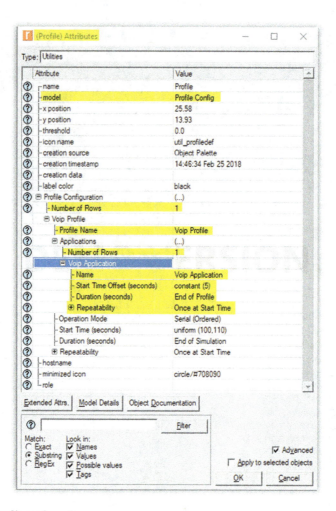

Fig. 1.3 Profile configuration

used for scenarios that are running GSM (as shown in Fig. 1.5). For the profile definition, the start time is set to 5-s offset for all scenarios (as shown in Fig. 1.6).

In order to simulate link failure and observe how the network recovers from it, the link between R1 and R2 is configured to process the failure recovery based on the information given in Table 1.2. The link fails and recovers after every 300-s. The simulation was running for 35 min. Figure 1.6 shows the Link Recovery configuration tab.

Before running our simulations, the DES statistics options and IP attributes were set for the appropriate protocol for each scenario. Depending on the considered scenario, the option "Sim Efficiency" was disabled for the RIP and OSPF so that the considered protocol can update its routing table when there is any change in the network such as link failure or recovery process.

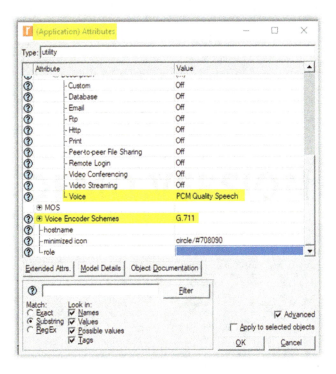

Fig. 1.4 G.711 setup

1.4.3 Routing Scenarios

For comparison purpose, four scenarios are configured with the network model shown in Fig. 1.1. Two of the scenarios are running the RIP protocol and the other two are running the OSPF protocol. One of the RIP (resp., OSPF) scenarios is running the G.711 codec and the other is running the GSM codec. Figure 1.7 depicts one of the RIP scenarios where all the network devices are configured to run the RIP, with the goal to compare the performance of the GSM and G.711 codecs. On the other hand, Fig. 1.8 depicts the same for one of the OSPF scenarios.

1.5 Simulation Results

In each graphic, the X-axis represents running time (minute) and the Y-axis represents the considered performance metrics being evaluated, which appear at the top (as a legend). For example, in Fig. 1.9, the Y-axis represents the *Voice Traffic Sent and Received* (in bytes/seconds).

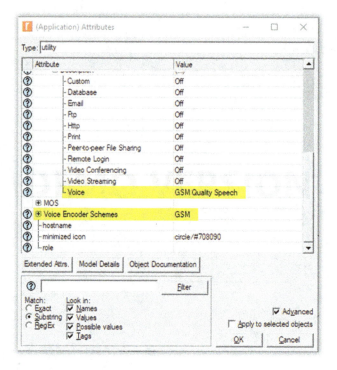

Fig. 1.5 GSM setup

1.5.1 Traffic Sent

This metric represents the total voice traffic sent from the source IP Phone to the destination IP Phone across the simulated network using the OSPF and RIP protocols. According to the results captured in Fig. 1.9, OSPF and RIP send about 6000 bytes/s using GSM and 30,000 bytes/s using G.711, which means that G.711 has sent 24,000 bytes more traffic per second than GSM at a given time.

1.5.2 Traffic Received

This metric represents the total voice traffic received from the source IP Phone to the destination IP Phone across the simulated network using the OSPF and RIP protocols. According to the results captured in Fig. 1.9, OSPF and RIP receive about 6000 bytes/s using GSM and about 30,000 bytes/s using G.711, meaning that G.711 receives 24,000 bytes more traffic per second than GSM at a given time, resulting in very low packet loss.

1 Performance Evaluation of G.711 and GSM Codecs on VoIP Applications... 9

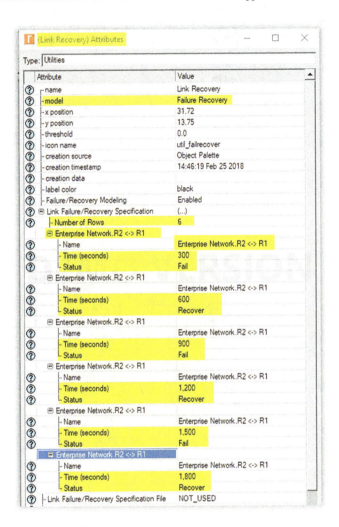

Fig. 1.6 Link recovery configuration

Table 1.2 Link recovery/failure

Time (s)	Status
300	Fail
600	Recover
900	Fail
1200	Recover
1500	Fail
1800	Recover

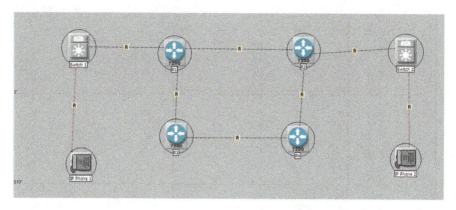

Fig. 1.7 RIP scenario

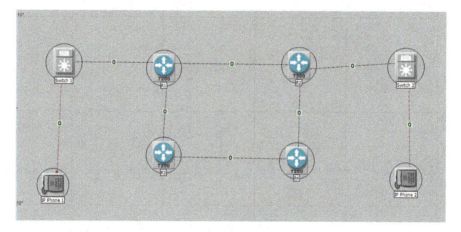

Fig. 1.8 OSPF scenario

1.5.3 Jitter

When packets are sent and received, there is an end-to-end delay variation between them. The variation in the arrival time of the packets at the receiver end leads to the jitter. The sender is expected to transmit each packet at a regular interval. Jitter can be caused by many factors such as congestion in the network, jitter buffer overload, timing drifting, or route changes. Ideally, a jitter of 0-ms is acceptable, but for higher-quality VoIP calls, jitter of 30-ms or less is also deemed acceptable [12].

1 Performance Evaluation of G.711 and GSM Codecs on VoIP Applications... 11

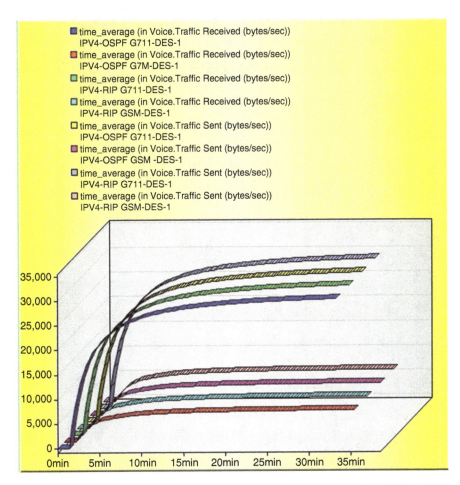

Fig. 1.9 Average voice traffic sent and received between GSM and G.711 using OSPF and RIP protocol

Based on the results captured in Fig. 1.10, OSPF and RIP have a similar performance for GSM and G.711 (jitter is less 0.00001-s), which means that a better quality of voice traffic is obtained since the lower the jitter, the better the quality of voice traffic.

1.5.4 End-to-End Delay

This metric represents the average time taken by voice packets to travel from source to destination. Delay can occur because of network congestion at the source or

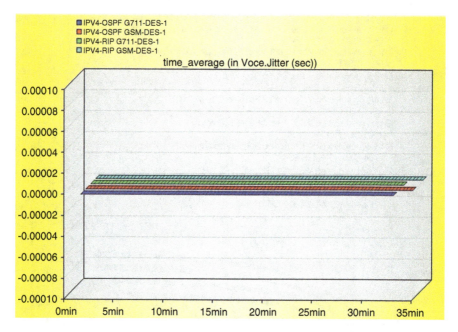

Fig. 1.10 Voice Jitter vs. time between GSM and G.711 using OSPF and RIP protocol

destination. For this metric, only the number of packets that have successfully been delivered to destination counts [13].

According to the results captured in Fig. 1.11, the end-to-end delay using GSM is 0.15-s and that using G.711 is 0.13-s for both OSPF and RIP, meaning that G.711 yields a lower end-to-end delay compared to GSM, resulting in a much faster voice encoding/decoding algorithm than that for GSM.

1.5.5 Packet Delay Variation

Packet delay variation (PDV) also known as IP Packet Delay variation is the difference in the end-to-end one-way delay between packets in transmission with any lost packets being ignored. The lower the packet delay variation, the better the performance is [13].

According to the results captured in Fig. 1.12, GSM has a PDV of about 0.000011-s whereas that of G.711 is about 0.000006-s for both OSPF and RIP,

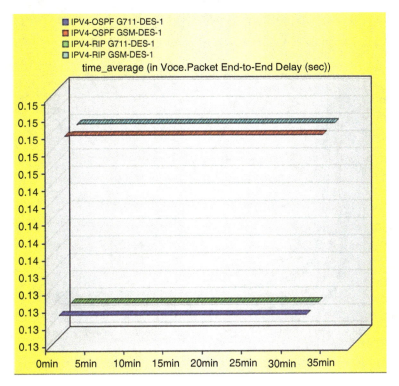

Fig. 1.11 End-to-end delay between GSM and G.711 using OSPF and RIP protocol

meaning that G.711 has a better voice encoding/decoding algorithm than GSM, which has led to a better quality of service for VoIP transmission compared to that of GSM.

1.5.6 Mean Opinion Score

The mean opinion score (MOS) is a subjective test [14], usually on a scale from 1 to 5, which does not have any unit. Here, a score of 5 means "excellent" and a score of 1 means "unacceptable." This metric is meant to measure the quality of experience of the user, especially for voice and video. It is used for individual opinion scores.

According to the results captured in Fig. 1.13, G.711 has a MOS score of 4.344 and GSM has a MOS score of 4.334 for both OSPF and RIP, i.e., the MOS score of G.711 is 0.01 higher than that of GSM, meaning that using G.711 results in a slightly better user experience than using GSM.

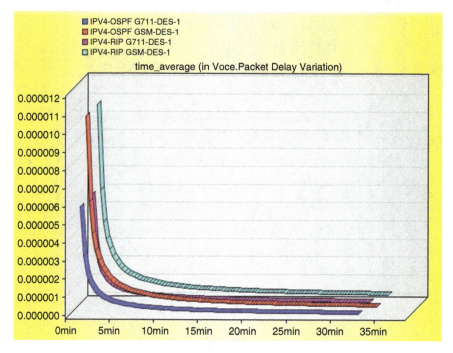

Fig. 1.12 Packet delay variation between GSM and G.711 using OSPF and RIP protocol

1.6 Conclusion

In this chapter, the performance of two different voice encoding codecs G.711 and GSM on VoIP applications using OSPF and RIP routing protocols have been conducted. Our simulation results have shown that the G.711 codec outperforms the GSM codec for all considered performance metrics, indicating that using the G.711 codec results in a better audio quality transmission for VoIP applications than using the GSM codec.

As future work, the FTP or Video server can be included to the studied topology and the performance of audio codecs on such topology can be investigated, using other routing protocols such as IS-IS. Other voice encoding codec such as G.723.1 or G.729A could also be analyzed, along with a comparison of their performance for IPv6 vs. IPv4.

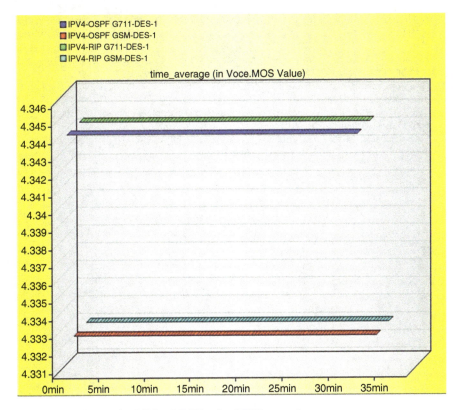

Fig. 1.13 MOS value for GSM and G.711 using OSPF protocol

References

1. M. Syed, I.Y. Ambore, Performance evaluation of OSPF and RIP on IPv4 and IPv6 technology using G.711 Codec. Int. J. Comput. Networks Commun. (IJCNC) **8**(6), 1–15 (2016)
2. Riverbed Modeler version 17.5, https://support.riverbed.com/content/support/software/steelcentral-npm/modeler-index.html. Last visited 6 July 2018
3. Voip-info.org, Codecs, https://www.voip-info.org/wiki/view/Codecs. Last visited 6 July 2018
4. CISCO, Configuring Voice over IP, https://www.cisco.com/c/en/us/td/docs/ios/12_2/voice/configuration/guide/fvvfax_c/vvfvoip.pdf. Last visited 11 Mar 2018
5. X. Che, L.J. Cobley, VoIP performance over different interior gateway protocols. Int. J. Commun. Netw. Secur. **1**(1), 34–41 (2009)
6. TechPublic, Understanding the protocols underlying dynamic routing, https://www.techrepublic.com/article/understanding-the-protocols-underlying-dynamic-routing/. Last visited 6 July 2018
7. H.P. Singh, S. Singh, J. Singh, S.A. Khan, VoIP: state of art for global connectivity-a critical review. J. Netw. Comput. Appl. **37**(1), 365–379 (2014)
8. W.M. Hussein, S. Jamwal, Comparative analysis of various routing protocols. Int. J. Mod. Eng. Res. **2016**, 2 (2016)
9. M. Ahmed, A.T. Litchfield, S. Ahmed, VoIP performance analysis over IPv4 and IPv6. Int. J. Comput. Netw. Inf. Secur. **11**, 43–48 (2014)

10. L. Besacier, C. Bergamini, D. Vaufreydaz, E. Castelli, The effect of speech and audio compression on speech recognition performance. in *IEEE 4th Workshop on Multimedia Signal Processing*, 2002, pp. 301–306
11. H. Sathu, M.A. Shah, Performance monitoring of VoIP with multiple codecs using IPv4 and IPv6to4 tunnelling mechanism on windows and Linux. Int. J. Model. Optim. **2**(3), 2–6 (2012)
12. How does internet work, what is jitter in networking? https://howdoesinternetwork.com/2013/jitter. Last visited 11 Mar 2018
13. CISCO, VoIP – an in-depth analysis, http://www.ciscopress.com/articles/article.asp?p=606583. Last visited 11 Mar 2018
14. VoIP mechanic, MOS - mean opinion score for VoIP, https://www.voipmechanic.com/mos-mean-opinion-score.htm. Last visited 11 Mar 2018

Chapter 2
Cyclic Redundancy Check Based Data Authentication in Opportunistic Networks

Megha Gupta

2.1 Introduction

Data authentication is the procedure to legitimize the source and trustworthiness of data. This property is must for communication within any type of network, whether it is wired, wireless, ad hoc, opportunistic, etc. Data authentication consists of the following two parts:

1. *Authentication*: To verify that the source (generator) of the message is a true and faithful entity.
2. *Integrity check*: To validate the nobility and fairness of the message sent.

In this paper, the discussion and proposed work is for data authentication technique in opportunistic networks (OppNets). An opportunistic network shown in Fig. 2.1 has main characteristic of highly mobile working conditions with unreliable wireless connectivity among the nodes. As shown in Fig. 2.1, vehicle, pedestrian, satellite, or any other available communication channel could be used to set up an opportunistic network. OppNets work on store and carry forward paradigm. These types of network are also identifying as delay tolerant networks because a participating node will buffer the data till it comes in contact with another node in the network. Hence data delivery may suffer high transmission delay.

Data security with authentication is an important aspect of communication in the OppNets [1–3]. Two majorly affected data authentication attacks are the following:

M. Gupta (✉)
Miranda House, University of Delhi, New Delhi, India

© Springer Nature Switzerland AG 2019
I. Woungang, S. K. Dhurandher (eds.), *2nd International Conference on Wireless Intelligent and Distributed Environment for Communication*, Lecture Notes on Data Engineering and Communications Technologies 27,
https://doi.org/10.1007/978-3-030-11437-4_2

Fig. 2.1 An opportunistic network

1. *Replay Attack*—Data packets are sending repeatedly or delayed by an evil node.
2. *Packet Alteration Attack*—Data is modified without the knowledge of source node.

Hence during message flow from source to destination, the replay prevention and the data packet validation is essential for secured message transportation. Data transmission from a source node to the destination may also be affected by security as well as non-security reasons. It may be difficult to predict the reasons of data packets lost. Resource limitations are one of the prime factors for packet lost. Another reason of legitimate packets lost may be due to the malicious attack. An evil node in the network may inject counterfeit packets to replace the legitimate data packets. This security breach raises a big question on the authentication of data received by the destination.

Message integrity and reliability must be maintained and the destination node should be able to check that message is not modified or changed with a fake message. For the above concerns, data encryption techniques could be used. Data encryption [4] method is to shield the original source data by jumbling it with some special keys. With this technique a malicious node in the network will be unable to view the source data, as it will not have the key to decrypt it. Decryption is converting back the data to original message by applying the key. Various data encryption methods are *cryptography, hashing, checksum, CRC*, etc.

The organization of the paper is as follows: Sect. 2.2 presents the related literature. Section 2.3 discusses the CRC methodology. Section 2.4 presents the proposed technique and Sect. 2.5 discusses the simulation analysis and results of this work. Conclusion and future work is in Sect. 2.6.

2.2 Related Work

All wireless communication systems need to provide security in the network due to ubiquitous nature of the transmission hence susceptible to various security attacks. Encryption and decryption methods are used to authenticate the data in any type of network. Various research works are in progress for security in OppNet. But the methods to provide data authentication and integrity in opportunistic networks are very few.

Kumar et al. [5] presented a security framework consisting of five properties for opportunistic networks. Author emphasized that authentication, privacy, trust, integrity, and non-repudiation are essential for secure communication in OppNets. In work [1] also, authors have provided a security framework for opportunistic networks. They discussed about authentication, application and user privacy, secure routing, co-operation, and trust.

Shifka et al. have worked on privacy in opportunistic networks. In their two different works [6] and [7], they have proposed two pairing based encryption techniques. The work [8] utilizes identity based cryptography for secure data transferring in intermittently connected mobile ad hoc networks.

In work [9], authors used a trust framework for establishing authentication in opportunistic networks. But it can be applied for limited environment and will work only if nodes are first registered with a super node. Ahmad et al. [2] have discussed about data authentication attacks in OppNets. They proposed a merkle tree based technique to provide security during communication within the network. But it requires more space to store the binary tree and its various levels. It may also become inefficient in terms of bandwidth as more data need to be transported along with the original message.

The works related to opportunistic network's data reliability are in small number, moreover the available reviews focus on user authentication rather than the correctness of communicated data. The proposed model using CRC provides an efficient technique in terms of lesser memory utilization, lesser calculations, and a better bandwidth utilization for maintaining data integrity.

2.3 Overview of CRC (Cyclic Redundancy Check)

CRC is a key which is appended with data to detect any modifications in data packets that may occur during transmission from source to destination. CRC uses a polynomial key to authorize the message. The same polynomial key is available to source and destination node. An example of the polynomial key is $x^2 + 1$ or in binary form, it is 101.

Source node generates encoded data with the use of data packet and the key. Key is generated by a Seed node, the node which is having authority to generate and distribute the polynomial keys to the other nodes of the network. CRC working at source and destination is explained below.

Generation of CRC Encoded Data by a Source Node

n: number of bits in data
k: number of bits in key

1. Data packet is first augmented with $k - 1$ zeros.
2. Divide the data packet by key (polynomial) using modulo-2 division.
3. rem: Store the remainder of the above division.
4. Encoded data will be: data + rem;
5. Encoded data packet will be used for transmission to destination.

Verification of Data by a Destination Node

1. Received data packet is divided again by the key (polynomial) using modulo-2 division.
2. If the remainder of the step 1 division is zero,
3. Then data is without any change, received successfully.
4. Else, data does not match with the original data, hence discard it.

2.4 Proposed Work

2.4.1 Assumptions

This work assumes that:

1. In OppNet, nodes can communicate with Wi-Fi/Bluetooth.
2. There is a Seed node, which is having low mobility and high pause time. It will be acting as CRC generator. And only Seed node is authorized to provide CRC to other nodes.
3. Malicious node can drop the original packet but could not alter the packet contents. It may also inject a new fake data packet in the network.

2.4.2 Authentication Key

This section describes the elements of an authentication key, which plays an important role in verification of the source message at the destination node.

Seed node generates an authentication key which consists of the following components as shown in Table 2.1:

1. *Timestamp, t*: it indicates the present timestamp.
2. *CRC key*: a polynomial identifier, e.g., $x^2 + 1$
3. *Validity time, t_v*: it indicates the time period for which the included CRC key is valid to authenticate the message.

2 Cyclic Redundancy Check Based Data Authentication in Opportunistic Networks

Table 2.1 Seed node authentication key

Timestamp	Key	Validity time
t	CRC value	t_v

Table 2.2 Seed node authentication key at time "t"

t		$x^2 + 1$		$t + 25$

Table 2.3 Seed node authentication key at time $t' + 20$

$t' + 20$		$2x + 1$		$t' + 15$

Table 2.4 Source packet description

Field name	Description
Source	Source node identity
Destination	Destination node identity
Time	Creation time of the message
Checksum	CRC checksum for the message
Message	Message of source

The three fields mentioned in the Seed node authentication key may be varied with time. To strengthen the authentication process, validity time may also be changed by Seed node for different CRC values as shown in Tables 2.2 and 2.3.

Tables 2.2 and 2.3 are presenting the example of authentication key generated by Seed node. At time t, message validating key is "$x^2 + 1$" and it is valid till next 25 s. At time t', message validating key is "$2x + 1$" and it is valid till next 15 s only.

Each node needs to receive CRC key from the Seed node, in order to send or forward message in the network. Nodes may be moving here and there in the network. On interaction with the Seed node, information about data authentication key is received by nodes from the Seed node. Any node will be able to send its message, if it appends CRC key received from Seed node in its data packet.

2.4.3 Algorithm

This section elaborates the working of the proposed technique. It provides two algorithms: one is to detect the fake data packets and another one is to check the integrity of message. Source node will create a data packet consists of fields as shown in Table 2.4.

Figure 2.2 is representing an OppNet environment where security breach will take place due to the presence of a malicious node. Node "$n1$" is the source node and node "$n4$" is the destination. There is one malicious node which may overhear the message in the network and inject the new fake data packet between the communications.

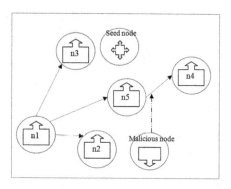

Fig. 2.2 Malicious node breaching security

During data authentication security breach, a malicious node can dribble one or more data packets but not majority portion of the message. If a source node sends 100 packets, then a malicious node could drop packets in range of 1–10% as all the packets will not be going through it only. In opportunistic network there is no dedicated path between source and destination. Message may be travelled different routes to reach destination due to high mobility in network.

Malicious node may also introduce fake packets to substitute the original data packets, and also stamp these lookalike data packets with current timestamp. Destination node will follow the steps mentioned in algorithm below to detect and drop the fake packets.

Algorithm 1 To detect fake data packets
---
1. read packetCreationTime from the received packet
2. $validtime = packetCreationTime[0]$
3. $validtimecounter = 1$;
4. j=0
5. For all packets
6. check packetCreationTime
7. $if packetCreationTime[i] == validtime$
8. $validtimecounter + +$;
9. $valid packet = true$;   proceed with step 5;
10. else
11.   do
12. $if packetCreationTime[i] == othertime[j]$
13. $othertimecounter[j] + +$;   proceed with step 5;
14.   while j!=0
15. $othertime[+ + j] = packetCreationTime[i]$
16. $othertimecounter[+ + j] + +$;
17.
18. $while j! = 0$
19. $check\ if\ othertimecounter[j] >= nbPktsThreshold$
20. $valid packet = true$
21. else
22.   this is a fake packet, drop it.

2 Cyclic Redundancy Check Based Data Authentication in Opportunistic Networks

---

Algorithm 2 To check the integrity of data

---
1. read packetCreationTime from the received packet
2. calculate the checksum of the packet from the CRC received by Seed node for that particular timestamp
3. compare the calculated checksum with the received checksum in the data packet.
4. if both checksum matches then
5.   the packet is intact.
6. else
7.   its a fake packet, drop it.

The data structure and variables used in the proposed algorithm are self-explanatory, e.g., *valid_time, valid_packet, packetCreationTime*, etc. Variable *nbPktsThreshold* is used to store the minimum number of packets with a timestamp that should reach the destination node. Data array *other_time* is used to cache the packet timings that are different from *valid_time*.

Destination node will perform the calculation to detect and drop the fake data packets. It will observe the packet's creation time. A message may consist of n data packets and all of them will be having same creation time.

The strategy is based on the facts that a malicious node could not drop all of the packets of a source. Assuming source "*A*" sends twenty packets for destination D and five packets were dropped by a malicious node. A malicious node also injects five new packets in the network. The source packets (fifteen undropped packets) will be having timestamp "*t*" and CRC1. And the new introduced five packets will be having different timestamp "*xt*" and may also have different CRC (as Seed node is changing it frequently) that is CRC2. And hence using the Algorithms 1 and 2, these malicious five packets could be easily identified and discarded by the destination node.

2.5 Results and Analysis

A scenario is implemented in the ONE simulator [10]. Routing protocol is taken as Epidemic. Nodes are communicating with Bluetooth having range of 10 m. Pedestrian speeds are between 1 and 1.5 m/s and vehicles are moving at speeds between 10 and 50 km/h. Three metrics are used to evaluate the proposed algorithm: percentage of authenticated delivery versus number of nodes, percentage of authenticated delivery versus number of malicious count, and percentage of altered/new packets detection.

For the purpose of comparison, results are taken once with Epidemic protocol and then with Epidemic protocol having proposed security algorithm implemented. Figure 2.3 is representing the percentage of authenticated delivery versus number of nodes. This metrics is calculated on the basis of number of original message packet received by the destination node. In the opportunistic network, the number

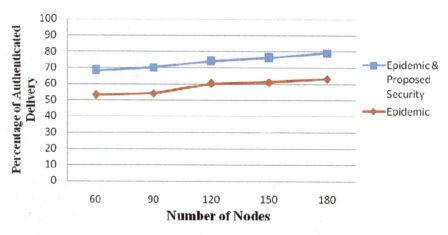

Fig. 2.3 Percentage of authenticated delivery vs number of nodes

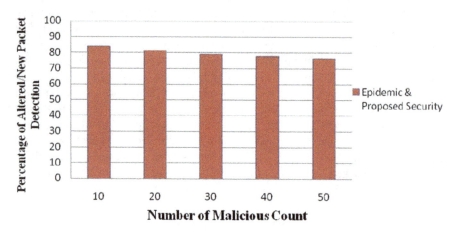

Fig. 2.4 Percentage of altered/new packet detection vs number of malicious count

of nodes is varied from 60 to 180. Due to mobility and scarcity of neighboring nodes, authenticated delivery percentage is approximately 58% in Epidemic protocol and 73% in Epidemic technique with the proposed security algorithm. The delivery percentage increases with the increase in number of nodes in the network as more than one path could be taken for successful delivery of source node message to destination.

The proposed method is productively able to detect the malicious attempts for faking the source packets as shown in Fig. 2.4. Percentage of altered/new packet detection is between 76 and 84%. This is due to transmission and computational delays. Even with the increase of malicious activities in the network, the proposed technique is isolating approximately 80% of the fake data packets.

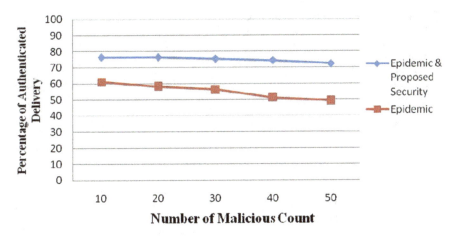

Fig. 2.5 Percentage of authenticated delivery vs number of malicious count

It can be easily inferred from Fig. 2.5 that the Epidemic protocol is achieving the high percentage of authenticated delivery with the proposed security feature. With the increase in number of malicious nodes in the network, the Epidemic protocol authenticated message delivery percentage decreases and falls to 49%. But with the provided security solution the authenticated message delivery percentage is approximate to 75%.

2.6 Conclusion

In opportunistic networks, routing with security is a challenging task due to the ubiquitous nature of nodes and network. In the proposed work, a better security technique has been provided for data authentication in the network. Using CRC method, message integrity is protected during communication between source and destination. Results are effectively showing that the proposed method is able to detect modification in the data packets sent by source node. In future, this technique will be amalgamated with an effective routing technique for OppNets. And the affect of the proposed technique over other routing protocols will also be observed.

References

1. Y. Wu, Y. Zhao, M. Riguidel, G. Wang, P. Yi, Security and trust management in opportunistic networks: a survey. J. Secur. Commun. Netw. **8**(9), 1812–1827 (2015)
2. A. Ahmad, R. Doss, M. Alajeely, K. Ahmad, Securing OppNets from packet integrity attacks using trust and reputation, in 2017 31st International Conference on Advanced Information Networking and Applications Workshops (WAINA) (IEEE, Piscataway, 2017), pp. 7–12

3. A. Shikfa, Security challenges in opportunistic communication, in IEEE GCC Conference and Exhibition (GCC) (IEEE, Piscataway, 2011), pp. 425–428
4. J. Wang, Z.A. Kissel, *Introduction to Network Security: Theory and Practice* (Wiley, Singapore, 2016)
5. P. Kumar, N. Chauhan, N. Chand, Security framework for opportunistic networks, in *Progress in Intelligent Computing Techniques: Theory, Practice, and Applications. Advances in Intelligent Systems and Computing*, vol. 719 (Springer, Singapore, 2018), pp. 465–471.
6. A. Shikfa, M. Onen, R. Molva, Privacy in content-based opportunistic networks, in IEEE International Conference on Advanced Information Networking and Applications Workshops (IEEE, Piscataway, 2009), pp. 832–837
7. A. Shikfa, M. Onen, R. Molva, Privacy in context-based and epidemic forwarding, in *IEEE International Symposium on a World of Wireless, Mobile and Multimedia Networks* (IEEE, Piscataway, 2009), pp. 1–7
8. Y. Ma, A. Jamalipour, Opportunistic node authentication in intermittently connected mobile ad hoc networks, in IEEE 16th Asia-Pacific Conference on Communications (APCC) (IEEE, Piscataway, 2010), pp. 453–457
9. U.P. Singh, N. Chauhan, Authentication using trust framework in opportunistic networks, in 2017 8th International Conference on Computing, Communication and Networking Technologies (ICCCNT) (IEEE, Piscataway, 2017), pp. 1–7
10. A. Keranen, J. Ott, T. Karkkainen, The ONE simulator for DTN protocol evaluation, in Proceedings of the 2nd International Conference on Simulation Tools and Techniques (ICST, Brussels, 2009)

Chapter 3
Hybrid Cryptographic Based Approach for Privacy Preservation in Location-Based Services

Ajaysinh Rathod and Vivaksha Jariwala

3.1 Introduction

The information society is mainly founded on the information and communications technologies (ICTs). The members of the information society want to obtain the information as fast as possible, from everywhere and at any time [1]. Emergency services, LBS reminder, map navigation, location-based marketing, location-based search, and location-based advertisement are examples of location-based services. LBSs are available on a variety of mobile platforms like mobile devices, PDAs, GPS devices, and other devices because they are ubiquitous. Nowadays, the growth of LBS users is very fast. In location-based application, users provide their highly personalized information like their identification information and location information to the service provider causing vulnerability to their privacy, e.g., an attacker can also get the current location of the user and also track user's daily activities. Due to the tracking capability, it opens many possibilities of computer based crimes like kidnapping, harassment, car theft, and many more. Varieties of attacks are already possible, so there is a big challenge to protect location privacy with minimum cost.

A. Rathod (✉)
Department of Computer Engineering, RDIC, C. U. Shah University, Wadhwan, Gujarat, India

V. Jariwala
Department of Information Technology, Sarvajanik College of Engineering and Technology, Surat, Gujarat, India

© Springer Nature Switzerland AG 2019
I. Woungang, S. K. Dhurandher (eds.), *2nd International Conference on Wireless Intelligent and Distributed Environment for Communication*, Lecture Notes on Data Engineering and Communications Technologies 27, https://doi.org/10.1007/978-3-030-11437-4_3

3.2 Literature Survey

In this section, we analyze popular category of crypto-based privacy model for location-based service that has been proposed by different authors.

3.2.1 Categorization of Crypto-Based Privacy Model for LBS

In the simple form of communication between an LBS user and LBS provider, the former sends a simple query (Q) containing an ID and his location (L) and a request for information (I) that he wants to retrieve from provider P [2]. A user provides his identity and location to provider, but provider is not always trust worthy.

Most schemas within this category adopt a centralized model for privacy [3]. There are many solutions that are already proposed by using TTP based schemas [4–19]. TTP based schemas are used very often and easy to deploy. This schema has many drawbacks so do not rely on TTP [1]. Many schemas are already proposed as TTP free schemas [20–24].

3.2.1.1 TTP Free Schema

Without the help of trusted third party, all users jointly compute the task which will improve the privacy of the users.

Collaborative-Based Schema

It is a fully distributed schema. The trust is scattered among the nodes that form an ad-hoc network. All peers work collaboratively as shown in Fig. 3.1 to achieve privacy among untrusted entities. Various algorithms are already proposed by Solanas et al. [7–9]. The advantage of this approach is that it does not rely on TTP, it is distributed and also guarantees us privacy, but it has issues with cost and scalability.

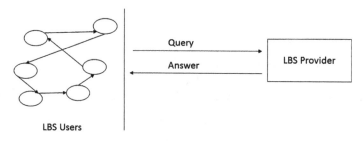

Fig. 3.1 The communication schema of collaborative method between users and LBS provider [3]

3.3 Motivation of Proposed Approach

Many schemas have been already proposed by different authors but they pose and open challenges that need to be solved, such as reducing communication cost, reducing computational cost, and poor scalability. Our aim is to propose a novel solution that provides location privacy to the LBS users. The main goal of the research is to achieve features such as TTP free, hybrid approach (centralized and decentralized), improve scalability, reduce cost in resource constraint devices, collision free, and enhance privacy.

Figure 3.2 represents the system architecture of proposed schema. It contains two main components: (1) LBS users and (2) LBS provider. Each user has their private information on their mobile like UserID U_{id} and location information (Lg_i, Lt_i). There is a need of preserving the privacy of LBS users. As our first step, we are finding the number of Users U_i in cloaking region who are requesting for location-based information. Next, we generate random region [25] R_i based on the clustering algorithms for users in spatial cloaking region. This procedure creates different cluster as shown in Fig. 3.2. All users will add some random value in their original location information because any malicious user can collude with location-based service provider. To avoid this attack, each user will add random value in their actual location by using secrete share function and perform a secure data aggregation using privacy homomorphism PH [10] in each random region R_i which uses centralized approach that is shown in Fig. 3.2 with blue edges. Next, we use the decentralized approach to perform random chaining RC for all distributed random region R_i to compute the secure centroid C as shown in Fig. 3.2 with green edges. The last user, U, sends the encrypted sum of location C to LBS provider P as shown in Fig. 3.2 with red edge. LBS provider will decrypt this sum of location by his own private key.

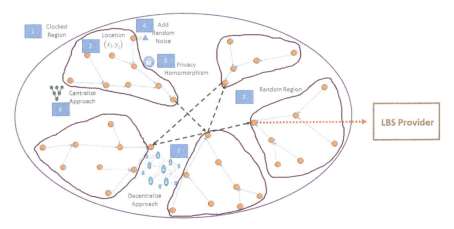

Fig. 3.2 Proposed communication schema of hybrid approach between users and LBS provider

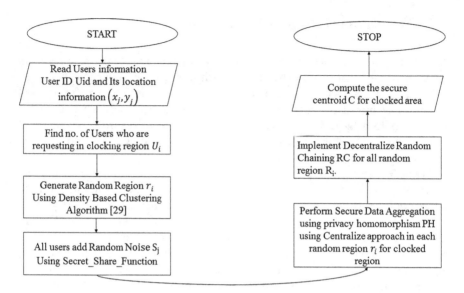

Fig. 3.3 Flowchart for proposed protocol schema

Algorithm 1 Users Communicate Using Proposed Model

Input: LBS Users U_i (User Identification U_{id}, Position information (Lg_i, Lt_i)).
Output: Compute Secure Centroid C.
1: Find LBS requesting Users in cloaking region U_i.
2: Create Region R_i using Clustering Algorithms in spatial cloaking region.
3: All users add random noise in their current location using Secret Share function.
4: Perform Secure Data Aggregation using privacy homomorphism [27, 28] PH using Centralized approach in each random region R_i for cloaked area.
5: Perform Decentralized Random Chaining RC for all random region R_i and compute the secure centroid C for cloaked area.
6: The Last User, U send encrypted sum C centroid to LBS provider P.
7: LBS Provider P perform decryption using his own private key and Find Centroid.

The main aim is to hide the user's location within the other users and also give inaccurate location information to the LBS provider. We use hybrid approach that includes distributed method to achieve minimum cost and improve scalability. In this paper, we propose a protocol schema that preserves privacy between users and LBS provider [26]. Proposed protocol schema is shown in Fig. 3.3.

3.4 Experimental Results and Evaluation

We have developed the simulation scenario and implemented the same in Java. We evaluated it on an Intel Core i3 2.30 GHz machine with 2 GB of RAM running Windows 7 OS. We experimented the performance with different density based clustering algorithm and different dataset of users. Performance metrics is measured in average computation time taken by the processes.

3.4.1 Datasets

In our simulation, we use dataset of Weeplaces,[1] which contain check-in activity of the users in location-based social network. It is also integrated with the API of other location-based social network (LSBN) like Facebook place (see footnote 1), Gowalla,[2] etc. Users have to login in location-based social network and they can connect with the other friends in this network, those who have already registered in this application. This dataset contains 7,658,368 check-ins generated by 15,799 users over 971,309 locations [2]. We use this dataset because users can connect with Weeplaces datasets and connect with their friends. They can also perform location-based search.

3.4.2 Density Based Clustering Algorithms

It is the process of making the groups of points together, which are close for the given dataset/set of points in space. This is called as density based clustering. Examples of density based clustering [29–31] are DBSCAN, OPTICS, etc. We use density based clustering algorithm DBSCAN and Optics to create random region in cloaked area.

3.4.3 Results

We have analyzed the performance of our model for various parameters like execution time and number of clusters based on various users as shown in Figs. 3.4 and 3.5 and Table 3.1. OPTICS algorithm gives better results as compared to DBSCAN clustering algorithm.

---

[1]http://www.yongliu.org/datasets/.

[2]http://techcrunch.com/2011/12/02/report-facebook-has-acquiredgowalla/.

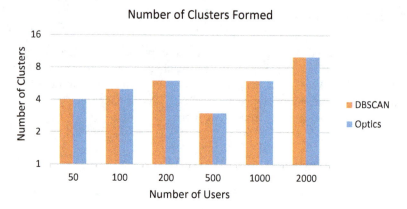

Fig. 3.4 Total execution time over number of users for Weeplaces dataset

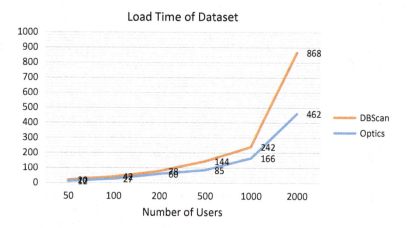

Fig. 3.5 Number of clusters over number of users for Weeplaces dataset

Table 3.1 Result of various parameters of Weeplaces dataset

No.	Number of users	Clustering algorithms	Total time (ms)	Number of clusters
1	50	DBSCAN	20	4
2	100		42	5
3	200		78	6
4	500		144	3
5	1000		242	6
6	2000		868	10
7	50	OPTICS	12	4
8	100		27	5
9	200		60	6
10	500		85	3
11	1000		166	6
12	2000		462	10

3.5 Conclusion

In location-based services, users will get some important information based on their location by providing their highly personalized data like user's identity, user's location information, etc. Privacy preservation is of paramount importance with the rapid growth of LBS users. In this paper, we address the benefits and issues of both TTP based and TTP free schema. In our proposed approach we select hybrid tech-nique to take advantage of centralized and centralized schema. This schema gives the guarantee of privacy of the users with improves scalability and reduces the cost. We have created random region from the given set of input dataset by using basic density based clustering algorithms—DBSCAN and OPTICS clustering algorithms. We have analyzed the performance of our model for various parameters like execution time and number of clusters based on various users. OPTICS algorithm gives 60% better result compared to DBSCAN. In future, we will perform homomorphic encryption, construct tree topology, and distributed random chaining approach as per our proposed schema.

References

1. R. Padmanaban, Location privacy in location based services: unsolved problem and challenge. Int. J. Adv. Remote Sens. GIS **2**(1), 398–404 (2013)
2. Y. Liu, W. Wei, A. Sun, C. Miao, Exploiting geographical neighborhood characteristics for location recommendation, in *Proceedings of the 23rd ACM International Conference on Information and Knowledge Management (CIKM'14)* (ACM, 2014), pp. 739–748
3. E. Magkos, Cryptographic approaches for privacy preservation in location-based services: a survey. Int. J. Inf. Technol. Syst. Approach **4**(2), 48–69 (2011)
4. M. Wernke, P. Skvortsov, F. Durr, K. Rothermel, A classification of location privacy attacks and approaches. Pers. Ubiquit. Comput. **18**(1), 163–175 (2014)
5. Y. Wang, F. Li, B. Xu, L2P2: Location-aware location privacy protection for location-based services, in *Proceedings - IEEE INFOCOM,* 2012, pp. 1996–2004
6. G. Yang, J. Li, S. Zhang, H. Zhou, A survey of location-based privacy preserving. J. Convergence Inf. Technol. **8**(11), 27–33 (2013)
7. A. Solanas, J. Domingo-Ferrer, A. Martinez-Ballest, Location privacy in location-based services: beyond TTP-based schemes, in *Proceedings of the 1st International Workshop on Privacy in Location-Based Applications,* October 2008
8. G. Ghinita, P. Kalnis, A. Khoshgozaran, C. Shahabi, K.-L. Tan, Private queries in location based services: anonymizers are not necessary, in *ACM SIGMOD International Conference on Management of Data,* 2008, pp. 121–132. ISBN: 978-1-60558-102-6
9. G. Ghinita, P. Kalnis, M. Kantarcioglu, E. Bertino, A hybrid technique for private location-based queries with database protection, in *Advances in Spatial and Temporal Databases Volume 5644 of the Series Lecture Notes in Computer Science* (Springer, Berlin, 2009), pp. 98–116. ISBN: 978-3-642-02982-0
10. A. Solanas, A. Martinez-Balleste, Privacy protection in location-based services through a public-key privacy homomorphism, in *Proceedings of the 4th European Conference on Public Key Infrastructure: Theory and Practice* (Springer, 2007). ISBN: 3-540-73407-4 978-3-540-73407-9

11. Y. Huang, R. Vishwanathan, Privacy preserving group nearest neighbor queries in location-based services using cryptographic techniques, in *IEEE Global Telecommunications Conference GLOBECOM*, 2010
12. R. Gupta, U.P. Rao, An exploration to location based service and its privacy preserving techniques: a survey. J. Wireless Pers. Commun. **96**(2), 1973–2007 (2017)
13. B. Amro, Y. Saygin, A. Levi, Enhancing privacy in collaborative traffic-monitoring systems using autonomous location update. IET Intell. Transp. Syst. **7**(4), 388–395 (2013)
14. M. Ashouri-Talouki, A. Baraani-Dastjerdi, Homomorphic encryption to preserve location privacy. Int. J. Secur. Appl. **6**(4), 183–189 (2012)
15. A.K. Tyagi, D.N. Sreenath, Preserving location privacy in location based services against Sybil attacks. Int. J. Secur. Appl. **9**(12), 175–196 (2015)
16. S. Patil, S. Ramayane, M. Jadhav, P. Pachorkar, Hiding user privacy in location base services through mobile collaboration: a review, in *International Conference on Computational Intelligence and Communication Networks IEEE*, 2015
17. T. Peng, Q. Liu, G. Wang, Enhanced location privacy preserving scheme in location-based services. IEEE Syst. J. **11**(99), 1–12 (2014)
18. R. Shokri, G. Theodorakopoulos, P. Papadimitratos, E. Kazemi, J.-P. Hubaux, Hiding in the mobile crowd: location privacy through collaboration, in *IEEE Transactions on Dependable and Secure Computing, Special Issue on "Security and Privacy in Mobile Platforms"*, 2014
19. A.K. Tyagi, N. Sreenath, Future challenging issues in location based services. Int. J. Comput. Appl. **114**(5), 51–56 (2015)
20. N. Yang, Y. Cao, Q. Liu, J. Zheng, A novel personalized TTP-free location privacy preserving method. Int. J. Secur. Appl. **8**(2), 388 (2014)
21. A. Solanas, A. Martinez-Balleste, A TTP-free protocol for location privacy in location-based services. Trans. Comput. Commun. **31**(6), 1181–1191 (2008)
22. C. Bettini, X. Sean Wang, S. Jajodia, Protecting privacy against location-based personal identification, in *Workshop on Secure Data Management SDM 2005: Secure Data Management*, 2006, pp. 185–199
23. G. Yang, J. Li, S. Zhang, H. Zhou, A survey of location-based privacy preserving. J. Convergence Inf. Technol. **8**(11), 27 (2013)
24. M. Wernke, P. Skvortsov, F. Durr, K. Rothermel, A classification of location privacy attacks and approaches. Pers. Ubiquit. Comput. **18**(1), 163–175 (2014)
25. S.R. Shastry, P.K. Deshmukh, A.B. Bagwan, Generating: random regions in Spatial cloaking algorithm for location privacy preservation. IOSR J. Comput. Eng. **9**(4), 46–49 (2013)
26. A. Rathod, V. Jariwala, Investigation of privacy issues in location-based services, in *Recent Findings in Intelligent Computing Techniques. Advances in Intelligent Systems and Computing*, vol. 707 (Springer, Singapore, 2018), pp. 55–65
27. V. Jariwala, D. Jinwala, Evaluating homomorphic encryption algorithms for privacy in wireless sensor network. Int. J. Adv. Comput. Technol. **3**(6), 1–11 (2011)
28. X. Zhu, Y. Lu, X. Zhu, S. Qiu, A location privacy-preserving protocol based on homomorphic encryption and key agreement, in *International Conference on Information Science and Cloud Computing Companion IEEE*, 2014
29. R.J. Patil, K.K. Joshi, S. Raksha, Analysis on preserving location privacy. Int. J. Adv. Res. Comput. Sci. Softw. Eng. **5**(3), 562–566 (2015)
30. J. Liu, J. Luo, J.Z. Huang, X. Li, *Privacy Preserving Distributed DBSCAN Clustering* (ACM, Berlin, 2012)
31. P. Batra Nagpal, P. Ahlawat Mann, Comparative study of density based clustering algorithms. Int. J. Comput. Appl. **27**(11), 421–435 (2011)

Chapter 4
Design of Energy-Aware PRoPHET and Spray-and-Wait Routing Protocols for Opportunistic Networks

Sibusiso Shabalala, Zelalem Shibeshi, and Khuram Khalid

4.1 Introduction

OppNets are designed to provide communication in challenging network environments where the network connection is intermittent and the message packet delivery is not guaranteed [1]. In such networks, nodes are highly mobile and due to the prevalence of opportunistic communication, the task of forwarding a message packet from source node to destination node is a challenge since there is no predefined end-to-end path. For this reason, the store-carry-and-forward mechanism is adopted to forward the message packets toward their destinations [2]. In this mechanism, the source node forwards the message packet to a certain number of relay nodes (intermediate nodes) that are in its direct communication range. Any selected relay node then stores the message packet in its buffer and awaits until it finds another suitable relay node which has the potential to forward the message packet to the destination or closer to the destination, hoping that the message packet will finally reach its end. As a result, every node in the network has an opportunity to be selected as a suitable relay node [3].

In OppNets, the next-hop selection process and the message packet forwarding mechanism are highly dependent on the selected routing protocol. There were, some challenges that these protocols tried to solve. First of all, forwarding the message packets without pre-existing knowledge of the relay nodes that might already have

---

S. Shabalala (✉) · Z. Shibeshi
Department of Computer Science, University of Fort Hare, Alice, South Africa
e-mail: 201406457@ufh.ac.za; zshibeshi@ufh.ac.za

K. Khalid (✉)
Department of Computer Science, Ryerson University, Toronto, ON, Canada
e-mail: khuram.khalid@ryerson.ca

© Springer Nature Switzerland AG 2019
I. Woungang, S. K. Dhurandher (eds.), *2nd International Conference on Wireless Intelligent and Distributed Environment for Communication*, Lecture Notes on Data Engineering and Communications Technologies 27,
https://doi.org/10.1007/978-3-030-11437-4_4

the message copies can lead to imperfect coordination among these nodes, causing multiple duplicate transmissions of the same message packet [4]. Another challenge is related with the node's storage. Every node is equipped with a limited storage; and keeping the message copy in the node's buffer for a longer period of time can lead to excessive unnecessary network resource consumption although the benefit of increasing the probability of successful delivery of the message to the destination prevails. Indeed, when a message has been successfully delivered to the destination, there are still many copies of it being transmitted by the relay nodes around the network [5]. This unnecessary transmission of already delivered messages consumes a lot of Transmit Energy when relay nodes send and receive these messages. Also, these additional copies occupy unnecessary spaces in the buffers of nodes and may cause congestion, which may lead to network performance degradation due to increased overhead ratio that may occur [5]. Developing energy-efficient routing protocols that can ensure high message delivery while keeping resource usage very low is one way of solving the abovementioned challenges.

This chapter proposes new energy-efficient versions of the PRoPHET [6] (referred to as EPRoPHET) and the Spray-and-Wait protocol (S&W) [7] (referred to as ES&W). The design of these protocols is based on the implementation of a one-hop acknowledgment mechanism that ensures that: (1) additional copies of the already delivered messages are removed from the buffers of nodes, and (2) the packets are transmitted only through those encountered relay nodes that have enough amount of remaining battery power; yielding some energy and network resources saving.

The remainder of this chapter is organized as follows. In Sect. 4.2, some related works are discussed. In Sect. 4.3, the design of the proposed routing protocols is presented. In Sect. 4.4, simulation results are given and discussed. Finally, Sect. 4.5 concludes the chapter.

4.2 Background and Related Work

In this section, we introduce the benchmark routing protocols PRoPHET and S&W, and we discuss few other energy-aware routing protocols for OppNets that have been investigated in the literature.

4.2.1 PRoPHET

Lindgren introduced PRoPHET [6], a probability-based routing protocol that maintains the probability of reaching the destination for each node. This protocol uses the history of previously encountered nodes and considers the transitive property inherent in human contacts to estimate the so-called delivery predictability (DP) of a node, a metric used to predict the probability for the relay node to successfully deliver the message to destination. PRoPHET is based on the idea that if a node has visited a location or contacted another node frequently, the chances to visit that location again in a near future are high. Based on this metric, a node forwards a message to the encountered relay node that has a higher DP value than its own DP value. This is meant to ensure that the message can only be forwarded to a more reliable relay node. The DP is calculated based on three parameters: direct probability, transitivity, and ageing.

The direct probability used to calculate the probability of transmitting the message directly between two nodes is obtained as [8]:

$$P\ (A, B) = P\ (A, B)_{old} + (1 - P\ (A, B)_{old}) \times P_{enc} \qquad (4.1)$$

where $P\ (A, B)$ is the delivery probability of node A to node B, $P\ (A, B)$old is the encounter probability of node A and B before the current encounter occurs, and P_{enc} is a scaling factor. Whenever node A (that has a message for destination D) encounters a relay node, node B, both nodes A and B will exchange their delivery predictability values. Next, node A will compare $P\ (A, D)$ against $P\ (B, D)$. If $P\ (B, D)$ is greater than $P\ (A, D)$, the message will be copied to node B, Otherwise, it will not be copied to node B and P_{enc} will be increased by 1. The following transitivity formula is used to calculate the probability of indirect contact of nodes through multiple-hop relays [8].

$$P\ (A, i) = P\ (A, i)_{old} + (1 - P\ (A, i)_{old})\ P\ (B, i)\ P\ (A, B) \times P_{enc} \qquad (4.2)$$

where $P\ (A, i)$ denotes the encounter probability between nodes A and i through relay node B, and it is obtained as follows [8]:

$$P\ (A, i) = P\ (A, i)_{old} \times \gamma^{T} \qquad (4.3)$$

where γ represents the ageing constant (its proposed value is 0.998) and T is the number of time units elapsed since the last time $P\ (A, i)$ was aged.

4.2.2 Spray-and-Wait

Spyropoulos et al. [7] proposed the S&W routing protocol, a two-phase's flooding-based mechanism meant to control the unnecessary replicas of the original message by limiting the number of copies that can be produced. The message is delivered in two phases, namely the spray phase and the wait phase. In the spray phase, the source node initially sprays n number of message replica to n number of distinct relay nodes. In the wait phase, if the destination node was not found at the spray phase, all n relay nodes having a copy of the message store the message on their buffers until they meet the destination node to perform the direct delivery of these message replicas [7].

4.2.3   Related Work

In the literature, few energy-aware routing protocols for OppNets have been investigated. In [9], Dhurandher et al. introduced an energy-aware routing scheme for OppNets that utilizes a genetic algorithm along with some known local heuristic information about the neighbouring group of nodes to decide on the best next hop of each source node. The paper states that fitness function that is used in genetic algorithms is utilized to develop the modified routing algorithm. On the other hand, Yao et al. [10] proposed another energy-aware routing scheme for sparse OppNets that relies on an asynchronous sleep approach. The authors implemented a strategy to isolate those nodes whose remaining energy is low, imposing them to enter a dormant status until a valid relay node toward the destination is found, and from that point on, the concerned node becomes awake. In [11], Chilipirea et al. introduced an energy-efficient BUBBLE Rap protocol [12], in which the decision by a node to select its best forwarder relies on a utility function that is introduced in [12]. The authors designed their protocol by including some energy-related parameters. By means of the utility function, a mechanism is implemented to identify only those intermediate relay nodes whose remaining energy level is sufficient enough to carry the message forward toward its destination.

4.3   Proposed EPRoPHET Routing Protocol

In this chapter, we propose and design two energy-aware versions of the well-known PRoPHET and S&W routing protocols for OppNets (referred to as EPRoPHET and ES&W).

4.3.1   EPRoPHET

EPRoPHET follows the same assumptions and design features as that of PRoPHET [6], but with the following implemented energy-aware mechanism:

- A minimum energy threshold/level is set by the network administrator to prevent an encountered relay node to be selected as a next-hop node if its energy level is found to be less than that of minimum threshold. In this chapter, the prescribed minimum threshold is 300 J.
- When a node, say n1, encounters any relay node, say n2, and n2 is the destination, and it is found that the current message (i.e. the message being sent) is not already delivered, n1 forwards that message directly to n2 according to the PRoPHET routing rules without the need to check the energy level of n2. However, if n2 is not a destination node and it is found that the current message is not already delivered, n1 checks the energy level of n2, and if that energy level is found to

be less than a prescribed threshold (set by the network system administrator), the message will not be sent to n2 and the search for a new relay node is launched again.

- When the message has been delivered to the destination node. The destination node sends acknowledgement message (Ack_M) containing a message ID, and source and destination node ID to the last encounter node that had sent the message to the destination. This process is called one-hop acknowledgment because only a one-hop Ack_M is sent from the destination to the last relay node, and the rest of the Ack_M flooding is done via an Ack_T information exchange among the other nodes [6]. After a destination node has sent an Ack_M to the last encountering node, both destination and last encountering node update their Ack_T and remove the message from their buffers to prevent further relaying of it. Once these two nodes update their Ack_T, they flood the Ack_T information with all the nodes they come in contact with to prevent them from further relaying the message. Those relay nodes will also update their Ack_T to remove the already delivered message from their buffer and floods their Ack_T to all the nodes they encounter until all copies of the message have been removed. The pseudocode of the EPRoPHET algorithm is shown in Algorithms 1, 2, and 3 in Fig. 4.1.

4.3.2 ES&W Routing Protocol

The ES&W routing protocol follows similar design steps like the EPRoPHET protocol, where the same one-hop acknowledgment mechanism has been implemented, but following the S&W protocol rules [7], yielding a pseudocode similar to that in Fig. 4.1, but with the Spray-and-Wait protocol substituting the PRoPHET protocol.

4.4 Performance Evaluation

In this section, the performance of our proposed EPRoPHET and ES&W protocols is evaluated using the ONE simulator [13] version 1.5.1 RC2, under the Short Path Map Base Movement model, and varying number of nodes.

4.4.1 Simulation Parameters

The nodes in ONE simulator [13] are positioned in a rectangular simulation area (4500 m $\times$ 3400 m), and their location is defined in the form of x and y coordinates within this area. By default, the nodes are mobile and are divided into six groups.

Step1: Select the next Neighbor Node (NN)
Step2: If NN is busy then go to Step1
Step3: Repeat all Message (M) of Current Node (CN)
      4a: if NN has M then go to Step3
      4b: check ACK_T (AT NN) of NN for M
         If ACK_T of NN has M then
         Remove M from buffer of CN
         Update ACK_T (AT CN) of CN
         goto Step3 for next M
         end if
      4c: If Energy Level of NN < Minimum Energy
         Threshold (MET) and NN is not
         Destination Node (DN) then
         goto Step3
    4d: If NN is DN then
      forward M to NN
      else Follow ProPHET to send message to NN

Algorithm 2 for the destination Node receiving a Message

Step1: Receive Message (M) from the Last Sender Node (LSN)
Step2: If Destination Node (DN) of M is Current Node (CN) then
    Send ACK_M to LSN
    update ACK_T of CN with ACK_M
    remove M from the buffer of CN
    end if

Algorithm 3 for the last sender receiving the acknowledgment

Step1: Receive Message (M) from the Destination Node (DN)
Step2: If M contains ACK_M then
    update ACK_T of CN with ACK_M
    remove M from the bu_er of CN
    end if

Fig. 4.1 Pseudocode of the EPRoPHET algorithm

The first and third groups consist of pedestrians, and they each have 40 nodes, with an average speed of 0.5–1.5 m/s. The second group consists of Cyclist with 40 nodes with an average speed of 2.7–13.9 m/s. The remaining three groups are Tram groups (representing cars), each with 2 nodes having an average speed of 7–10 m/s. In our simulations, the number of nodes is increased by a scale of 40 nodes (for the first 3 groups) until we reached 160 nodes.

The simulation parameters that are considered in our simulations are captured in Table 4.1. The main parameters are explained as follows: (1) Initial Energy is the total energy of the nodes before the simulation starts, (2) Scan Energy is the

4 Design of Energy-Aware PRoPHET and Spray-and-Wait Routing Protocols... 41

Table 4.1 Simulation parameters

Parameters	Values
Simulation area	4500 m × 3400 m
Movement model	Short path map base movement model
Buffer size	5 M
Transmission range	10 m
Transmit speed	250 K
Message size	500 K, 1 M
Routing protocols	ES&W, S&W, EPRoPHET, and PRoPHET
Simulation time	43,200 s
Communication interface	Bluetooth
Initial energy	5000 J
Scan energy	0.1 J
Transmit energy	0.2 J
Scan response energy	0.1 J
Base energy	J

energy consumed when nodes discover each other, (3) Transmit Energy is the energy consumed by each node in sending the message to another node, (4) Scan Response Energy is the amount of energy consumed for the device discovery response, and (5) Base Energy is the amount of energy consumed when the node is not performing any transmission and is idle.

4.4.2 Performance Matrix

In our simulations, the following performance metrics are considered:

Average Remaining Energy: this is a total average remaining energy of the nodes at the end of the simulations.

Delivery Ratio (DR): this is a total number of delivered messages (denoted by D) to the destination node over the total number of created messages (denoted by C) at the source nodes, i.e. DR = D/C.

Overhead Ratio: it is the average number of relayed copies per message, and it is calculated as (number of relayed messages − number of delivered messages)/(number of delivered messages).

Dead Nodes: nodes whose remaining energy is less than 300 J (half the prescribed energy level) at the end of simulation.

4.5 Simulation Results

4.5.1 Comparison of EPRoPHET and PRoPHET

In this simulation, the number of nodes is varied and the impact of this variation on the average remaining energy, number of delivered packets, number of dead nodes, and overhead ratio is investigated. The results are captured in Fig. 4.2a–d, respectively. In Fig. 4.2a–d it can be observed that EPRoPHET outperform PRoPHET intering of average remaining energy number of dead nodes, overhead ratio and number of delivered messages to the destination node.

In Fig. 4.2a, it is observed that as the number of nodes is increased, the average remaining energy decreases, which is due to the fact that, there is a high interaction between nodes. This has also led to a high energy consumption due to the scan energy and scan response energy. For instance, when a node, say node A, discovers another node, say node B, then node A consumes a specific amount of energy (so-called scan energy), and when the discovered node (here node B) responds to node A, it also consumes some amount of energy. The higher the number of nodes available in the network, the higher the DP of nodes. This means that there is high possibility of nodes to often meet each other. As a result, there is a high number of messages transmitted and received in the network, which in turn results to a high energy

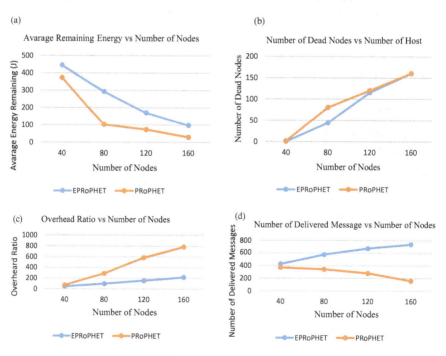

Fig. 4.2 EPRoPHET vs. PRoPHET

consumed by nodes each time they send or receive a message copy. Figure 4.2a also shows that EPRoPHET yields a higher energy level than PRoPHET. With the number of nodes kept to 40, EPRoPHET save more than 19% energy as compared to PRoPHET, as the number of nodes increased to 80, EPRoPHET save about 187.57% more energy as compared to PRoPHET. This is due to the fact that EPRoPHET ensures a low energy usage compared to PRoPHET by preventing those nodes that have lower energy level than the prescribed threshold to be selected as next hop. On the other hand, Fig. 4.2d also shows that as the number of nodes increases, the number of delivered messages also increases for EPRoPHET, but decreases for PRoPHET. This is attributed to the fact that in the case of PRoPHET, when a message reaches its destination, there are still many copies of it circulating in the network, and this unnecessary transmission of additional copies of already delivered messages consumes a lot of energy in the sending and receiving of messages, which can lead to a higher number of dead nodes as shown in Fig. 4.2b. Also, these copies occupy unnecessary spaces in the nodes' buffers and may degrade the network performance due to increased overhead ratio. Therefore, EPRoPHET reduces the network overhead and energy consumption by avoiding unnecessary transmission of the same message already delivered to its destination due to its intrinsic one-hop acknowledgment mechanism.

4.5.2 Comparison of S&W and ES&W Routing Protocol

In this simulation, the number of nodes is also varied and the impact of this variation on the average remaining energy, number of delivered packets, and overhead ratio is investigated. The results are captured in Fig. 4.3a–c, respectively. In Fig. 4.3a–c, it can be observed that as the number of nodes increases, the number of messages delivered to their destination and the average remaining energy of nodes also increase. This is due to the fact that both S&W and ES&W save energy and ensure low overhead ratio by limiting the number of message copies initially spread to the network by the source node during the spray phase. Furthermore, if the message copy does not reach the destination node during the spray phase, the relay nodes having a copy of the message will have to perform direct delivery to the destination node. In this way, the nodes in the S&W and ES&W protocols preserve a lot of energy (here transmit energy) that should have been depleted during the transmission stage, as the results there were no dead nodes for both ES&W and S&W routing protocol throughout the simulation.

Another interesting observation that can be seen from Fig. 4.3a is that the average remaining energy of S&W and ES&W are almost the same due to the reasons mentioned earlier. The total saved energy by ES&W throughout the simulation is less than 2%. However another interesting observation from Fig. 4.3c, is that ES&W yields a lower overhead ratio compared to that generated by S&W. This

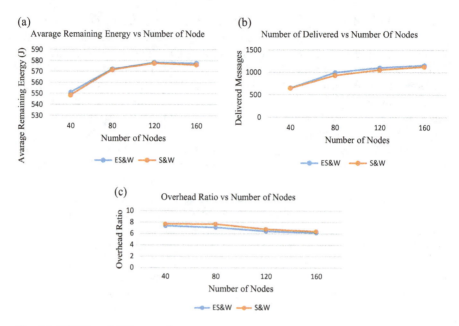

Fig. 4.3 ES&W vs. S&W protocol

is attributed to that fact that ES&W reduces the resource usage by removing the copies of the already delivered messages in the network using its intrinsic one-hop acknowledgment mechanism which avoids future transmission of these messages in the network. This process ensures low network overhead ratio, low energy usage, and high delivery ratio of messages to destinations. In general, removing unnecessary message copies on the buffer of a node allows it to have more memory space to carry undelivered messages to their destinations for a longer time, which increases the chance for the messages to reach their destinations.

4.5.3 Comparison of ES&W and EPRoPHET Routing Protocol

In this simulation, the number of nodes is also varied and the impact of this variation on the average remaining energy, number of delivered packets, and overhead ratio is investigated. The results are captured in Fig. 4.4a–c, respectively. In Fig. 4.4a–c, it is observed that compared to EPRoPHET, ES&W saves more energy and yields a higher delivery ratio and low overhead ratio. This is attributed to the same reasons already discussed above.

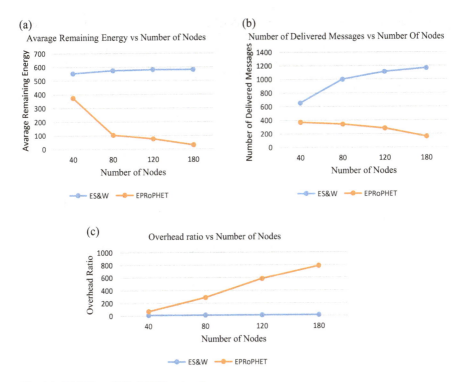

Fig. 4.4 ES&W vs. EPRoPHET protocol

4.6 Conclusion

In this chapter, two energy-efficient routing protocols for OppNets have been proposed, namely ES&W and EPRoPHET. Simulation results have shown that: (1) ES&W (resp., EPRoPHET) enhances the performance of S&W (resp., PRoPHET) protocol in terms of energy consumption, resource utilization, and number of messages delivered to destination; and (2) compared to EPRoPHET, ES&W saves more energy and yields a higher delivery ratio and low overhead ratio. As future work, we plan to evaluate the proposed energy-efficient routing protocols against other benchmark routing protocols for delay tolerant networks. We also plan to simulate the proposed ES&W and EPRoPHET protocols using real mobility traces.

References

1. R.S. Vaishali, C. Manicka, Disruption Tolerant Network (DTM), its network characteristics and core application. Int. J. Comput. Sci. Mob. Comput. **2**(9), 256–262 (2013).
2. D. Huang, Z. Gao, K. Niu, X. Qiu, Adaptable quota-stretchy routing for delay-tolerant networks. Proc. Comput. Sci. **107**, 513–519 (2017)

3. L. Pelusi, A. Passarella, M. Conti, Opportunistic networking: data forwarding in disconnected mobile ad hoc networks. IEEE Commun. Mag. **44**(11), 134–141 (2006)
4. V. Anshul, S. Anurag, Integrated routing protocol for opportunistic networks. Int. J. Adv. Comput. Sci. Appl. **2**(3), 85–92 (2011)
5. K. Khalid, *A History-Based Energy-Efficient Routing Protocol for Opportunistic Networks*, vol. 2 (Memorial University of Newfoundland, Toronto, 2014), pp. 85–92.
6. S. Grasic, E. Davies, A. Lindgren, A. Doria, The evolution of a DTN routing protocol RoPHETv2, in *6th ACM Workshop on Challenged Networks*, Las Vegas, 2011.
7. T. Spyropoulos, K. Psounis, C.S. Raghavendra, Spray-and-Wait: An efficient routing scheme for intermittently connected mobile networks, in *Proceedings of the SIGCOMM Workshop on Delay-Tolerant Networking*, pp. 252–259, , 22–26 August 2005
8. A. Lindgren, A. Doria, Probabilistic routing in intermittently connected network. Mob. Comput. Commun. Rev. **7**(3), 19–20 (2013)
9. S.S. Dhurandher, K.S. Deepak, W. Isaac, G. Rohan, G. Sanjay, GAER: genetic algorithm-based energy-efficient routing protocol for infrastructure-less opportunistic networks. J. Supercomputing **69**(3), 1183–1214 (2014)
10. Y.K. Yao, W.H. Liu, W.X. Zheng, Z. Ren, An energy-saving routing algorithm for opportunistic networks based on asynchronous sleep approach. Appl. Mech. Mater. **441**(3), 1001–1004 (2014)
11. A.C. Chilipirea, A. Petre, Energy-aware social-based routing in opportunistic network, in *Proceedings of the 27th IEEE WAINA*, pp. 791–796, 25 March 2013
12. P. Hui, J. Crowcroft, E. Yoneki, Bubble Rap: Social-based forwarding in delay tolerant network, in *Proceedings of the 9th ACM International Symposium on Mobile Ad Hoc Networks and Computing*, New York, 2008.
13. A. Keränen, J. Ott, T. Kärkkäinen, The ONE simulator for DTN protocol evaluation, in *Proceedings of the Second International ICST Conference on Simulation Tools and Techniques* (2009), pp. 1–10.

Chapter 5
An Asymmetric RSA-Based Security Approach for Opportunistic IoT

Nisha Kandhoul and Sanjay Kumar Dhurandher

5.1 Introduction

The internet of things (IoT) [1] is an integration of a number of technologies and different objects around us that are capable of interaction and cooperation with neighboring things to achieve some common goals. With the rapid developments in advanced cellular networks, big data, and social networks, the applications of internet of things (IoT) have attracted attention from both the academia and industries. Under IoT network's vision, the future internet will nurture the interactions between human beings, human-based communities and smart things held by them in a harmonious manner.

In terms of topology, the network connections are divided into following types: infrastructure-based and infrastructure-less connection. Infrastructure-based connections use preexisting infrastructure and manage the information in a centralized way. Infrastructure-less networks are based on opportunistic networks and use short-range radio techniques like RFID, Bluetooth, etc. to form noncentralized, ad hoc networks without the use of any infrastructure. Opportunistic networks are human centric as these networks use the opportunistic contact nature of human beings.

Opportunistic IoT (OppIoT) [2] extends the concepts of Opportunistic networks by merging human users and their smart devices. Mobility and short-range sensing technologies are used for collection, routing, and sharing of data among objects and human-based communities in OppIoTs.

N. Kandhoul (✉)
CAITFS, Division of Information Technology NSIT, University of Delhi, New Delhi, India

S. K. Dhurandher
CAITFS, Department of Information Technology, Netaji Subhas University of Technology, New Delhi, India

© Springer Nature Switzerland AG 2019
I. Woungang, S. K. Dhurandher (eds.), *2nd International Conference on Wireless Intelligent and Distributed Environment for Communication*, Lecture Notes on Data Engineering and Communications Technologies 27,
https://doi.org/10.1007/978-3-030-11437-4_5

The excessive heterogeneity of devices and vast scale of OppIoT systems magnifies the security threats. The conventional security means are not applicable to OppIoT technologies as they have limited computing power and scalability issues as large number of devices are interconnected. Users' data are insecure as they are exposed to attacks from uncertified users or devices present across the network. In OppIoT-based systems, securing the routing procedure [3] is a challenging task as the packet travels across varying network topology and uniformity must be maintained while routing the packets between source and destination. Thus, security of the routing process needs significant research contributions.

The security requirements for context- and content-based routing are divergent. The context-based routing techniques disclose the true identity of the nodes for performing routing. So, the identity of the nodes needs to be secured for such routing techniques. For content-based routing, the messages exchanged must be protected. When the messages exchanged across the network are read by unintended receivers, it is called eavesdropping attack [4]. The eavesdropping can act against the privacy of users when the messages carry the control information about the configuration of the network. So, the need of the hour is to design efficient security techniques for communications across such networks.

Since opportunistic networks are a class of opportunistic IoT systems and the opportunistic use of IoT devices is also feasible in condition where devices presence is uncertain, this chapter proposes a lightweight content-based security scheme for OppNets (called asymmetric RSA-based security approach for opportunistic IoT (RSASec)) that can be used in OppIoT scenarios. The designed scheme uses RSA-based asymmetric cryptography [5] to secure the messages and performs routing by predicting the node's future location using Markov chain [6] and considering node's probability of moving towards the destination. The encryption key is public, shared with all objects present in system, whereas the decryption key is kept secret (private). RSA algorithm is secure as it is very difficult to factor huge integers which are the multiplication of two huge prime numbers.

This chapter is composed of several sections. Section 5.2 provides the details of the related work carried out in this field. Section 5.3 illustrates the proposed work and the algorithms used in detail. Section 5.4 details the simulation results and the suggested approach's performance is evaluated. Finally, Sect. 5.5 concludes the work and highlights the future work.

5.2 Literature Review

Various security and privacy-based schemes for routing have been proposed in the literature for IoT and OppNets, but no significant work has been done for OppIoT. This section provides an overview of several secure routing protocols for IoT and OppNets.

A secure multihop routing protocol (SMRP) [7] was given by Chze and Leong allowing the IoT devices to communicate securely. Authentication of IoT devices is done before joining or creating a new network. The trust-aware secure routing framework (TSRF) [8] designed for WSNs was given by Hummen et al. where the trust is derived by making direct and indirect observations of behavior of the sensor nodes. However, TSRF uses huge amount of memory for complex trust calculations among nodes. Two-way authentication security scheme having its base on RSA algorithm and tailored over low-power wireless personal area networks (6LoWPANs) [9] for IPv6 was proposed for IoT using existing standard for internet, datagram transport layer security (DTLS) protocol.

The information security of the network concentrates on the data confidentiality and integrity, and user authentication. Encryption methods can be used to provide the security. Key-based cryptography approaches can be divided into symmetric key cryptography, public key cryptography, and hybrid key cryptography. Public key cryptography techniques were first applied to resource-limited sensors based on elliptic curve cryptography by Malan et al. [10]. This opened the door for the creation of several cryptographic mechanisms using public key for sensors. Access control mechanism was given by Hengartner et al. [11] using identity-based cryptography. Oliveira et al. [12] proposed a cryptography library for resource-limited sensors based on pairing. The authentication and access control mechanism given in [13] established the session key based on elliptic curve cryptography (ECC).

The approaches to control access are classified based on the type of policy: policy-based, attribute-based, and group-based access control. An authentication solution was proposed by Graffi et al. [14]. Fine-grained access control on data was provided by the use of a trust-based infrastructure. The messages were encrypted using the receiving node's public key, and intermediate nodes are allowed to acquire and create copies of these data, but are unable to read the data. Only the privileged users are able to decrypt and read the data. Two-channel cryptography was proposed by Jia et al. [15]. As the DTN nodes come in geographical range of each other, manual channel is used for the transmission of the verification data and the traditional wireless channel performs public key exchange.

A privacy preserving and personal profile matching scheme is where a pseudo-ID and a private key based on its attributes were assigned to users in the secure system-based handshake [16]. Ding et al. [17] designed an anonymity-based authentication and protocol for key agreement for DTNs. Defrawy et al. [18] focused on secure establishment of the context for DTNs. The users use their social contacts for the exchange of private messages. An approach for confidentiality of node's location was given by Gongjun et al. [19] for the vehicular networks.

Intrusion detection approaches were proposed for protecting unauthorized access to the network. These schemes are divided into three classes: signature-based, specification-based, and anomaly detection. Wang et al. [20] gave a cooperative intrusion detection architecture using a detection engine utilizing the analysis of the social networks. Qinghua et al. [21] countered the flooding-based attacks for DTNs by fixing a limit for message generation and its replication. Any attempt to breach this established rate limit is detected using a distributed scheme.

5.3 Proposed Work

5.3.1 A Secure Location Prediction-Based Forwarding Scheme for Opportunistic Internet of Things

The proposed RSASec protocol assumes an opportunistic IoT environment composed of n mobile nodes during the message forwarding process. The nodes are assumed to have sufficient buffer capacity to save the context information and each node has n neighboring nodes that participate in the message transmission. RSASec scheme uses RSA [5] based asymmetric cryptography to secure the network. The routing is performed considering node's probability of moving towards the destination which is predicted using Markov chain based on past location and direction of movement of a node.

Asymmetric cryptography makes use of public and private key, to encrypt and decrypt the messages, respectively. The public key is shared with every member node of network, whereas the private key is kept secret by the receiving node. The source node makes use of public key for message encryption, and the private key is stored secretly at the destination and is used to decrypt the message once delivered at the destination. The intermediate nodes can replicate and forward the messages but are unable to read the contents as it can be decrypted using only the private key of destination. RSA algorithm assures the integrity, confidentiality, nonreputability, and authenticity of messages.

The process of generating the keys is the most complex part of RSA cryptography. Miller–Rabin primality test (https://en.wikipedia.org/wiki/Miller) is used to generate two very large prime numbers, p and q. A modulus n is computed by taking the product of p and q. This number is used for generating the public and private keys. The length of modulus n converted into bits gives the length of the key. The public key is composed of the computed modulus n and an exponent, e. The value of e is set as the prime number that is not too large. The value of e is not secret as the public key is shared with all the things in the OppIoT system. The private key is composed of the modulus n and the private exponent d. Extended Euclidean algorithm (https://www.rsa.com/) is used to compute the multiplicative inverse with respect to the totient of n. Figure 5.1 summarizes the RSA encryption used for routing. The algorithm for generating keys used for encryption and decryption is shown in Algorithm 1. Once the keys are generated, they are saved for future use.

Whenever a sender wishes to send some message to destination, it encrypts the message with the public key generated during the key generation phase. Encryption process is described in Algorithm 2. The public key is used for the encryption of the message. Then, the routing Algorithm 3 is called to perform the message routing.

RSASec comprises of two phases, first the node's location/region is predicted using Markov chain and past location and direction of movement. Then, node's delivery probability is calculated using the predicted outputs. This procedure is described in Algorithm 3.

5 An Asymmetric RSA-Based Security Approach for Opportunistic IoT

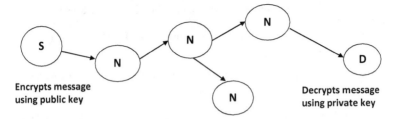

Fig. 5.1 Secure RSA-based routing. Where S source node, N intermediate nodes, and D destination node

---

Algorithm 1 Generation of Keys

1: **Begin**
2: Select two large primes randomly, p and q, and keep them secret
3: Value of n is calculated as:
$n = p * q$
4: Value of $\varphi(n)$ is computed as:
$\varphi(n) = (p-1)*(q-1)$
5: Random integer e is selected so that $1 < e < n$ and $\gcd(e, \varphi) = 1$
6: Value of d is computed so that $1 < d < \varphi(n)$ and $e * d = 1 \bmod \varphi(n)$
7: The public key thus generated is (e,n)
8: The private key thus generated is (d,n)
9: Save the generated keys
10: **End**

Algorithm 2 Encryption

1: **Begin**
2: Sender generates message m
3: Read public key
4: Encrypt the message as:
$em = m^e \bmod n$
5: Call Algorithm 3 for performing routing
6: **End**

Algorithm 3 RSASec

1: **Begin**
2: Initialize the matrix of transition probability for source $(T\ Ps)$ and neighbor $(T\ Pn)$
3: **for** every messages in the node buffer **do**
4:   **for** each n **do**
5:     Find the region and state of each n (i.e., $S0, S1, S2, S3,$ and $S4$)
6:     Calculate the probability that n moves from state 1 to another state (i.e., for the node in $S0$ to move to another state with probability $PS00, PS01, PS02, PS03,$ and $PS04$)
7:     Update the matrix of transition probability for node n:
$T\ Pn = (T\ Pprevious + Pi,j)\ /\ 2$

8: Calculate the final matrix of transition probability for n with the help of source/carrier node. $T\,Pnew = (T\,Pn \times T\,Ps)$

9: Predict the next region/location of node n using a Markov chain $T\,Pn,next = [nv0, nv1, nv2, nv3, nv4] \times T\,Pnew$

10: **end for**

11: Predict the next region/location of the source/carrier node using a Markov chain:
$T\,PS,next = [Sv0, Sv1, Sv2, Sv3, Sv4] \times T\,Pnew$

12: **for** each n **do**

13: Calculate the delivery probability of message (DP) of the node n from the predicted values, i.e., $T\,Pn,next$ using the weighted matrix:
$P\,rob_n = T\,P_{n,next} \times [W_0, W_1, W_2, W_3, W_4]_{(5,1)}$

14: **end for**

15: Calculate the delivery probability of message (DP) of the source/carrier node S from the predicted values, i.e., $T\,PS,next$ using the weighted matrix:
$P\,rob_S = T\,P_{S,next} \times [W_0, W_1, W_2, W_3, W_4]_{(5,1)}$

16: **end for**

17: **for** each neighboring node n of the source/carrier **do**

18: **if** $P\,rob_n \geq P\,rob_S$ **then**

19: Node n is inserted in Hashmap

20: **end if**

21: **end for**

22: **for** every node n in Hashmap **do**

23: Messages from source/carrier are transferred to the node n in Hashmap

24: **if** n is the destination node **then**

25: Call Algorithm 4 for performing Decryption

26: **end if**

27: **end for**

28: **End**

Algorithm 4 Decryption

1: **Begin**

2: Read private key

3: Perform decryption as:
$m = em^e \bmod n$

4: **End**

The intermediate nodes are able to perform the routing, take decisions, and calculate the delivery probability of the message, without being able to read the contents of the message. Only the destination has the rights to decrypt and read the message contents. Once the message reaches its intended destination, decryption is performed using the private key of the destination. This process is described in Algorithm 4.

5.4 Simulation Results

RSASec protocol's performance is computed for metrics like delivery probability of the node, average number of dropped messages, and average latency (in seconds). The results obtained are then compared to the Epidemic [22], Prophet [23], and

5 An Asymmetric RSA-Based Security Approach for Opportunistic IoT

Table 5.1 Parameters used for simulation

	Parameter	Value
1	Area of simulation	4500×3400 m
2	Node count	96
3	Group of nodes	6
4	Pedestrian group count	3
5	Nodes in each pedestrian group	30
6	Pedestrian's walking speed	0.5–1.5 km/h
7	Buffer size of pedestrian	15 Mb
8	Tram group's count	3
9	Number of nodes in each tram group	2
10	Tram speed	6.5 km/h
11	Tram buffer size	50 Mb
12	Transmission speed of Bluetooth	250 Kb/s
13	Transmission range of Bluetooth	20 m
14	Transmission speed of high-speed interface	10 Mb/s
15	Transmission range of high-speed interface	1500 m
16	Time-to-live of message	100 min
17	Interval of message generation	25–35 s
18	Size of message	500 Kb to 1 Mb
19	Movement model of nodes	Shortest path map-based movement model
20	Time of simulation	10,000 s

LPFR-MC protocols, under varying time-to-live (TTL) of message, total number of nodes in OppIoT, and the generation interval of messages. The simulation of the given approach is done using the opportunistic network environment (ONE) simulator [24].

The parameters considered for performing the simulation are given in Table 5.1. Mobility of the nodes is decided by movement models such as map-based movement model, random way point, shortest path map based, etc. In this work, shortest path map-based mobility model is used for performing simulations as it yields best results.

This work handles message modification attack. The malicious nodes modify the message content by appending some random text. The number of malicious nodes in the network is varied from 5 to 25% and the effect is noted on the packet delivery. The number of modified packets received on average in RSASec is 19.9% lower as compared to LPFR-MC. Also, the correct packets received are higher by 19.74% in RSASec. The results obtained are shown in Fig. 5.2.

The count of nodes present in the OppIoT are varied in the range 66–186, and the impact of this change is investigated against the performance metrics. The results obtained are given in Figs. 5.3, 5.4, and 5.5.

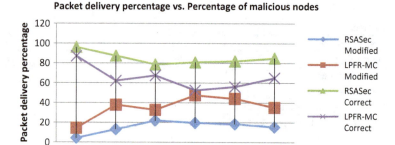

Fig. 5.2 Packet delivery percentage vs percentage of malicious nodes

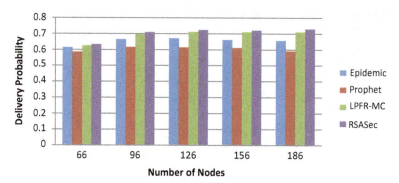

Fig. 5.3 Delivery probability vs. number of nodes

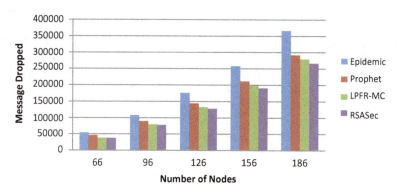

Fig. 5.4 Number of messages dropped vs. number of nodes

Figure 5.3 shows that as the number of nodes present in OppIoT network is increased, RSASec yields the largest delivery probability of 0.7. The average delivery probability for RSASec is 16.53% superior to the delivery probability of Prophet routing protocol, 7.54% superior than the value obtained for Epidemic, and 1.8% superior to LPFR-MC.

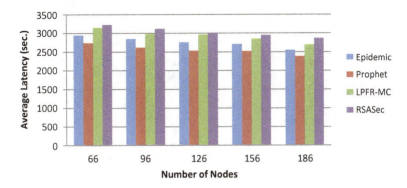

Fig. 5.5 Average latency vs. number of nodes

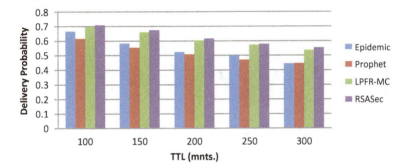

Fig. 5.6 Delivery probability vs. TTL

Figure 5.4 shows the impact of changing the count of nodes in OppIoT network on the count of messages dropped. The number of messages dropped rise with the rise in the number of nodes. The number of messages generated in OppIoT system rises with an increase in the number of nodes, resulting in further message drops. The average count of dropped messages for RSASec is observed to be 139,654 which is the lowest. The average number of messages dropped for RSASec is 11.76% superior to Prophet, 37.47% superior to Epidemic, and 3.9% superior to LPFR-MC.

Figure 5.5 shows that raising the number of nodes reduces the average latency. This is because a rise in the node count raises the count of message copies in OppIoT network thereby helping in speedy delivery of the message, thus lowering the average latency. It is noted that the average latency observed for RSASec is 3032.2 s, which is the lowest. The average latency for message delivery by RSASec is 18.72% superior to Prophet, 9.89% superior to Epidemic, and 3.56% superior to the value obtained using LPFR-MC.

The time-to-live (TTL) of the messages is ranged from 100 to 300 min, and the effect of this change is studied on the abovementioned performance metrics. The results thus obtained are given in Figs. 5.6, 5.7, and 5.8.

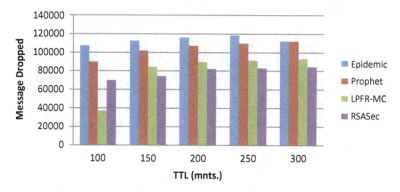

Fig. 5.7 Number of messages dropped vs. TTL

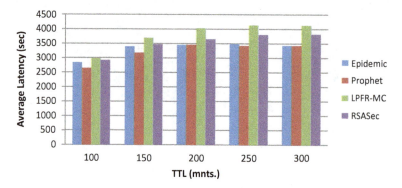

Fig. 5.8 Average latency vs TTL

Figure 5.6 shows that the delivery probability reduces when the TTL increases. Raising the time-to-live for messages increases the lifespan of the messages, resulting in a rise in the count of messages saved in the buffer. This increased message count in buffer leads to a rise in the messages dropped, thus reducing the message delivery probability. The average delivery probability of RSASec is 0.626, and this value is 20.92% superior to Prophet, 15.44% superior to Epidemic, and 2.1% superior to the value calculated for LPFR-MC.

Figure 5.7 displays that the count of dropped messages increases with the increasing value of the message TTL. The lifespan of the message rises with an increased TTL, resulting in the rise of messages dropped in the system. The average count of messages dropped is lowest for RSASec which is 78,804. The average number of messages dropped with an increasing TTL for RSASec is 32.11% lower than Prophet, 43.91% lower than Epidemic, and 0.28% lower than LPFR-MC.

Figure 5.8 conveys that the total delay in message delivery, i.e., average latency for all protocols under investigation rises with a rise in the message TTL. The message lifespan increases with a rise in the TTL leading to an increased delay in message delivery. Average latency for RSASec is 3537.44 which is 9.57% higher

5 An Asymmetric RSA-Based Security Approach for Opportunistic IoT

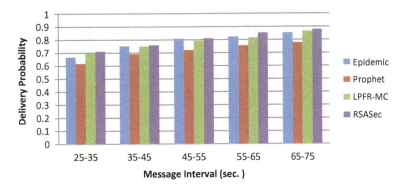

Fig. 5.9 Delivery probability vs message generation interval

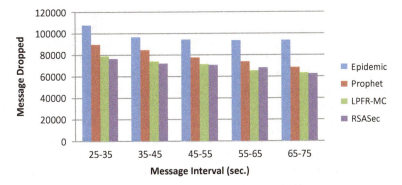

Fig. 5.10 Number of messages dropped vs. message generation interval

than that obtained using Prophet, 6.39% higher than Epidemic, and 6.76% lower than that obtained using LPFR-MC, respectively.

Lastly, the interval for generation of messages is varied in the range from 25–35 s to 65–75 s. The effect of the message interval change is noted against the metrics of performance, and the results thus obtained are displayed in Figs. 5.9, 5.10, and 5.11.

Figure 5.9 shows that the delivery probability of a message rises with increasing message generation interval. The RSASec algorithm outperforms all the investigated protocols in terms of probability of message delivery. The delivery probability for RSASec is 0.8 that is 12.7% higher than that obtained using Prophet, 2.8% higher than Epidemic, and 2.33% higher than that obtained using LPFR-MC, respectively.

From Fig. 5.10, it is clear that increasing the interval of message generation reduces the count of dropped messages for investigated protocols. The count of messages dropped for RSASec is lowest, i.e., 69,786. The average count of dropped messages for RSASec is 12.6% lower as compared to Prophet, 39.09% lower than Epidemic, and 0.92% lower than that obtained using LPFR-MC, respectively.

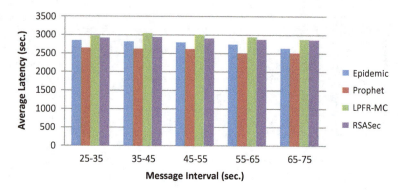

Fig. 5.11 Average latency vs. message generation interval

Figure 5.11 shows that the average delay in the delivery of messages rises with a rise in the interval of message generation. The average latency for RSASec is noted to be 2898.72 which is 10.9% higher as compared to Prophet, 4.45% higher than Epidemic, and 2.46% lower than that obtained using LPFR-MC, respectively.

The above graphs show that the suggested security scheme yields very fine results for all the performance metrics. This approach adds a little overhead as compared to the basic LPFR-MC protocol but improves the overall security with high delivery probability, low latency, and low count dropped messages in OppIoT network. The user's security is enhanced and as the messages exchanged are encrypted, the user's data is protected from eavesdropping attack and other cryptography-based attacks. Also, RSA-based asymmetric cryptography provides additional security as the keys are not to be exchanged between sender and receiver; they both compute their separate keys.

5.5 Conclusion

The suggested approach presents a basic secure routing protocol for OppIoT called asymmetric RSA-based security (RSASec) approach. The proposed RSASec scheme uses RSA-based asymmetric cryptography to secure the network and uses Markov chain to make prediction about a node's future location and its corresponding delivery probability. The results obtained by simulation prove that the proposed RSASec is superior in performance as compared to the existing routing protocols and LPFR-MC in terms of delivery probability of message, count of messages dropped, and average latency. The suggested approach protects the user from eavesdropping attack and other cryptographic attacks. In future, we plan to compare the performance of this approach to trust-based security schemes.

Also, we plan to implement a trust and cryptography-based security mechanism and to compare its effectiveness against the current work.

References

1. L. Atzori, A. Iera, G. Morabito, The internet of things: a survey. Comput. Netw. **54**(15), 2787–2805 (2010)
2. B. Guo, D. Zhang, Z. Wang, Z. Yu, X. Zhou, Opportunistic IoT: exploring the harmonious interaction between human and the internet of things. J. Netw. Comput. Appl. **36**(6), 1531–1539 (2013)
3. D. Airehrour, J. Gutierrez, S.K. Ray, Secure routing for internet of things: a survey. J. Netw. Comput. Appl. **66**, 198–213 (2016)
4. Y. Zhang, S. Yuanyu, et al., On secure wireless communications for IoT under eaves-dropper collusion. IEEE Trans. Autom. Sci. Eng. **13**(3), 1281–1293 (2016)
5. X.-J. Lin, L. Sun, H. Qu, An efficient RSA-based certificateless public key encryption scheme. Discrete Appl. Math. **241**, 39–47 (2018)
6. S.K. Dhurandher, S.J. Borah, I. Woungang, A. Bansal, A. Gupta, A location prediction-based routing scheme for opportunistic networks in an IoT scenario, in *Proceedings of Elsevier Journal of Parallel and Distributed Computing*, 2017
7. P.L.R. Chze, K.S. Leong, A secure multi-hop routing for IoT communication, in *Internet of Things (WF-IoT), 2014 IEEE World Forum on* (IEEE, 2014)
8. X. Anita, J. Martin Leo Manickam, M.A. Bhagyaveni, Two-way acknowledgment-based trust framework for wireless sensor networks. Int. J. Distrib. Sens. Netw. **9**(5), 952905 (2013)
9. J. Montavont, D. Roth, T. Nol, Mobile IPv6 in internet of things: analysis, experimentations and optimizations. Ad Hoc Netw. **14**, 1525 (2014)
10. D.J. Malan, M. Welsh, M.D. Smith, A public-key infrastructure for key distribution in TinyOS based on elliptic curve cryptography, in *Sensor and Ad Hoc Communications and Networks* (IEEE, 2004), p. 7180
11. U. Hengartner, P. Steenkiste, Exploiting hierarchical identity-based encryption for access control to pervasive computing information, in *SECURECOMM*, 2005, p. 384396
12. L.B. Oliveira, M. Scott, J. Lopez, R. Dahab, TinyPBC: pairings for authenticated identity-based non-interactive key distribution in sensor networks, in *INSS*, 2008, p. 173180
13. N. Ye, Y. Zhu, R.-C.B. Wang, R. Malekian, Q.-M. Lin, An efficient authentication and access control scheme for perception layer of internet of things. Appl. Math. Inf. Sci. **8**(4), 16171624 (2014)
14. K. Graffi, P. Mukherjee, B. Menges, D. Hartung, A. Kovacevic, R. Steinmetz, Practical security in P2P-based social networks, in *Proceeding of the IEEE 34th Conference Local Computer Networks*, 2009, p. 269272
15. Z. Jia, X. Lin, S.-H. Tan, L. Li, Y. Yang, Public key distribution scheme for delay tolerant networks based on two-channel cryptography. J. Netw. Comput. Appl. **35**(3), 905913 (2012)
16. R. Lu, X. Lin, X. Liang, X. Shen, A secure handshake scheme with symptoms-matching for mHealthcare social network. J. Mobile Netw. Appl. **16**(6), 683694 (2011)
17. Y. Ding, X.-W. Zhou, Z.-M. Cheng, W.-L. Zeng, Efficient authentication and key agreement protocol with anonymity for delay tolerant networks. Wireless Pers. Commun. **70**(4), 14731485 (2013)
18. K. El Defrawy, J. Solis, G. Tsudik, Leveraging social contacts for message confidentiality in delay tolerant networks. in *Proceedings of the 33rd Annual IEEE International Computer Software Applications Conference*, 2009, p. 271279
19. Y. Gongjun, S. Olariu, M.C. Weigle, Providing location security in vehicular ad hoc networks. IEEE Wireless Commun. **16**(6), 4855 (2009)

20. W. Wang, H. Man, Y. Liu, A framework for intrusion detection systems by social network analysis methods in ad hoc networks. Security Commun. Netw. **2**(6), 669685 (2009)
21. L. Qinghua, Z. Sencun, C. Guohong, Routing in socially selfish delay tolerant networks, in *Proceeding of the IEEE INFOCOM*, 2010, p. 19
22. L. Pelusi, A. Passarella, M. Conti, Opportunistic networking: data forwarding in disconnected mobile ad hoc networks. IEEE Commun. Mag. **44**(11), 134–141 (2006)
23. A. Lindgren, A. Doria, D. Schelen, Probabilistic routing in intermittently connected networks, in *Proceeding of ACM SIGMOBILE Mobile Comp. Commun.*, 2003, pp. 19–20
24. A. Keranen, J. Ott, T. Karkkainen, The ONE simulator for DTN protocol evaluation, in *Proceeding of 2nd Intl. Conference on Simulation Tools and Techniques (SIMU-Tools' 09)*, Rome, Italy, Mar 2–6, 2009, pp. 1–9

Chapter 6
Performance Analysis of A*-Based Hop Selection Technique in Opportunistic Networks Through Movement Mobility Models

Pragya Kuchhal and Satbir Jain

6.1 Introduction

Opportunistic network (OPPNET) [1] is the subclass of delay tolerant networks (DTN) [2, 3]. OPPNET can be deployed in the environment where no path exists for most of the time. This network is sparse and dynamic in nature just like MANET [4], but the data transmission process is carried out differently in the OPPNET. The store-carry-and-forward paradigm is the ruler in this network. This mechanism helps to communicate messages in the network. Data packets are exchanged only when the node lies within the same radio range or whenever they get the right opportunity to be paired, else the node stores the data packets in its buffer and waits for a better opportunity. The transmission of the messages/data packets is carried through the process known as routing. Designing of imperative routing protocols in OPPNETs is not easy and in addition to this to test the performance of routing protocol [5] in real environment is even more challenging. That's why most of the routing protocols are executed on simulators based on different mobility movement models [6, 7]. The rest of the chapter is formulated as follows: Section 6.2 presents the brief overview of already designed routing protocol A*OR and standard mobility models [8] in OPPNETs. Section 6.3 discusses the experimental environment [9]. Section 6.4 presents the analysis of the results obtained. And, Sect. 6.5 summarizes the chapter.

P. Kuchhal (✉) · S. Jain
Netaji Subhas Institute of Technology, University of Delhi, New Delhi, India

© Springer Nature Switzerland AG 2019
I. Woungang, S. K. Dhurandher (eds.), *2nd International Conference on Wireless Intelligent and Distributed Environment for Communication*, Lecture Notes on Data Engineering and Communications Technologies 27,
https://doi.org/10.1007/978-3-030-11437-4_6

6.2 Related Work and Background

Here in this section, a brief overview of A*OR [10] is discussed. Moreover, a brief description of mobility models is also discussed, namely the random way-point (RWP) [6], real time traces (RTT) [11], and the shortest path map-based mobility model (SPMBMM) [7].

6.2.1 A* Search-Based Next Hop Selection Routing Protocol: A*OR

This routing is one of the latest protocols in infrastructure-less opportunistic network. A*OR [10] predicts the neighboring nodes which are closer to destination by dynamically calculating the Euclidean distance of those nodes whose distance is less than the distance between the source and destination. In addition to the distance, the direction is also determined in order to strengthen the node properties and to smoothen the transmission process of messages in the network. After this, it compares the value of angle formed by the direction of nodes known as (θ) with a threshold say T_θ. Further, the delivery predictability of nodes is traced out to find more suitable node for transmitting the data packet in the network. The A* search algorithm helps in locating the best and worthy nodes in the network. This scheme evaluates the fitness value of the nodes by estimating the delivery predictability of the source node to neighbor/relay node and by heuristic value with the help of Eq. (6.1).

$$P_f = P_i * P_h \tag{6.1}$$

where:

- P_f = Total fitness value between source and destination.
- P_i = The delivery predictability between source and intermediate node n.
- P_h = The heuristic value of neighbor node to final destination.

Nodes which have the fitness value greater than certain fitness threshold value are considered as the best nodes for further transmitting the data packets to their desired destination.

6.2.2 Random Way-Point Model

RWP [6] is the basic and simplest model that is frequently used for simulating the settings in ONE simulator. Under this model, the nodes move randomly in any direction without any constraints implying on them. RWP mobility model has a random pause time once it finishes each movement segment in the random walk. In this model, direction angles, speeds, and pause times are sampled from a uniform distribution. During simulation, the simulation area is fixed, over which the random coordinates are assigned to the nodes. The node resides at a certain location for a specific duration of time and then proceeds towards the new destination with an arbitrary speed. This process continues throughout the simulation, and the nodes run along these zigzag paths that are already defined in the map of Helsinki city based on RWP model. This model is easy to implement and can be mathematically determined.

6.2.3 Shortest Path Map-Based Movement Model

In map-based mobility model (MBM) [7], node's movement is based on map data, that is, it has the road on the map to be chased by the nodes in that particular simulation area. This movement model randomly determines the direction of the nodes and restricts the mobility area geographically. This mobility model provides the special privilege of bifurcating the nodes into the groups depending on the area/map of map used by the groups. Map data is interpreted in well-known text (WKT) and MBM easily understands this data. For simulation purpose, the nodes take their position randomly on the simulation area/map and move on a path continuously until and unless they touch the end of the road or approach the intersection. As and if the end of road is approached, the node returns back and if somehow the node is on intersection, then a new path is selected other than the previous path that is through which it has come from. In this chapter, the improvised version of MBM, i.e., SPMBMM, is considered for simulation. SPMBMM is shortest path map-based mobility model which uses Dijkstra's shortest path algorithm for the movement of nodes on the map. Initially, the nodes take their position randomly on the map and move ahead towards the destination in the map following the Dijkstra's algorithm. Once the node reaches the destination, it has to wait for a moment and then has to select a new destination. In the map, all points have equal probability to be the destination node for better model. But, some unique method or point of interests are inosculated in map data so as to select the best and worthy destination node for transmitting the data packets in the network. The unique method or point on interest could be the special spots where the node is compelled to stop for a moment like bus stand, colleges, etc.

6.2.4 Real Time Traces

The proposed network model is modelled as a set of mobile nodes in which nodes join and leave the network at any time. It consists of one group of moving element (nodes) that is of pedestrian. Real encounter trace for the ONE simulator is haggle/info comm 2006 [11]. Nodes in a group follows the motion based on the stationary movement model. The elements in motion can be varied, which has no effect on the characteristics of communication. In this framework, the total nodes are 98. All the nodes utilizes the Bluetooth interface with a transmission speed of 2 Mbps and a range of 10 m.

6.3 Experimental Setup

6.3.1 Experiment Performance Metric Setup

1. **Delivery Predictability:** Defined as the probability of total number of messages successfully received by the sink within a given instant of time.
2. **Throughput:** It is defined as the ratio of the number of messages conveyed to destination and number of messages generated by the source node.
3. **Overhead ratio:** It is also defined as the average number of replicas transmitted per message and represented as:

$$Overhead\ Ratio\ (OR) = \frac{Relayed\ Messages - Delivered\ Messages}{Number\ of\ Total\ Delivered\ Messages}$$

4. **Average Latency:** It is determined as the average delay in message from message generation to the final delivery.
5. **Average Buffer Time:** It is illustrated as the average time calculated for each node's message residency in the buffer.
6. **Average Hop Count:** It is determined by the average of number of relay nodes navigated by a data packet to grasp its desired destination.

6.3.2 Experimental Environment

The performance of A*OR is studied by Opportunistic Network Environment (ONE) [12, 13] simulator using several mobility models. ONE simulator imports the data and visualizes the node movement using a GUI which helps in validating the model. $4500 \times 3400\ m^2$ of world size is considered for tracing the node's movement. Simulation runs for 100,000 s, and every message is accredited with 100 min TTL. There is one pedestrians group with 98 nodes (users) walking with the speed of [0.5, 1.5]m/s consists of specific buffer size of 15MB. Bluetooth communication speed is 250 KBps with a range of 20 m. The high-speed interface transmission speed is 10 MBps, with a range of 1500 m. A new message is provoked in every [25, 35] s with the size of [500 KB to 1 MB].

6.4 Experiment Result Analysis

The performance metric is evaluated while varying the TTL and Message Generation Interval with respect to delivery predictability, hop count, number of messages dropped, average latency of the messages, and the overhead ratio.

Figures 6.1, 6.2, 6.3, 6.4, and 6.5 show the behavior of the mobility model when the TTL varies with respect to abovesaid metric. It is observed from Fig. 6.1 that the delivery probability is very high for SPMBMM as in comparison to RWP and RTT when the TTL varies in the network. The average delivery predictability for SPMBMM is 0.68854, for RWP it is 0.18516, and for RTT it is 0.11492. The standard deviation defined for RWP, RTT, and SPMBMM is 0.064556846, 0.051910284, and 0.026204015, respectively.

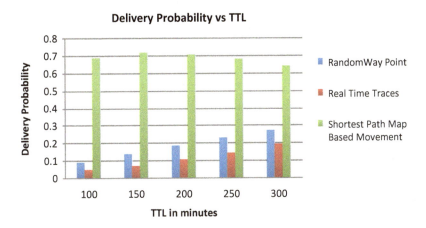

Fig. 6.1 Delivery probability versus TTL

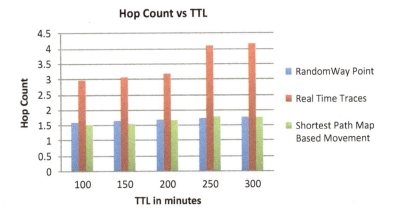

Fig. 6.2 Hop count versus TTL

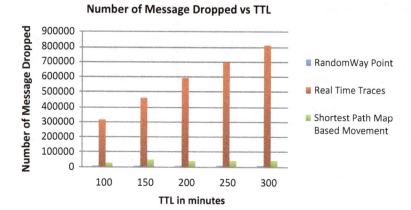

Fig. 6.3 Dropped messages versus TTL

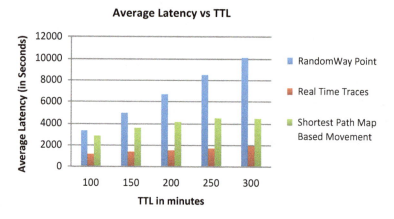

Fig. 6.4 Average latency versus TTL

Figure 6.2 shows that SPMBMM needs lower number of average hops for communicating the message in the network. However, it is still comparable for both RTT and RWP at higher values of TTL. But in case of RTT, very high number of hops are required for transmitting the message in the network as the TTL increases gradually. The average hop count for RWP is 1.68782, RTT is 3.49924, and SPMBMM is 1.6588. The standard deviation traced when the TTL varies is as follows: 0.061225661 for RWP, 0.524804854 for RTT, and 0.106688125 for SPMBMM.

Figure 6.3 discusses about the drop rate of messages in the network with respect to TTL variation. The figure clearly shows that the RWP has dribbled/dropped minimum number of messages while conveying between source and destination. Whereas this drop is highest in RTT which results in low delivery predictability under this scenario. On an average, 8114 is the message drop in RWP, 640178 for

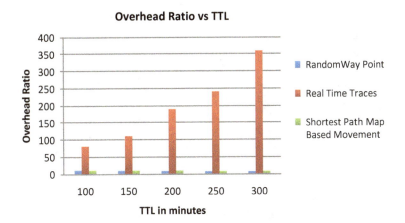

Fig. 6.5 Overhead ratio versus TTL

RTT, and 40177 for SPMBMM. 906.7839875 is the standard deviation for RWP, 129819.6609 for RTT, and 7100 for SPMBMM.

From Fig. 6.4, variation in delay of messages is observed with respect to gradual increase in TTL. It is figured out from the graph that RTT has minimum average delay as in comparison to rest of two RWP and SPMBMM. The average latency is calculated for SPMBMM, RWP, and RTT, and found that RTT has least average latency and SPMBMM has second minimum average latency that is 1543.98586 and 3906.44466, respectively. RWP has maximum average delay time while communicating the message in the network that is 6713.30684. The standard deviation reviewed for different mobility models is as follows: 2409.375624 for RWP, 269.0010442 for RTT, and 616.0693387 for SPMBMM.

Figure 6.5 discusses about the overhead ratio of the protocol in the network with respect to variation in the TTL. From the graph, it is analyzed that the SPMBMM has minimum overhead ratio as compared to RWP and RTT. On an average, the overhead ratio calculated for SPMBMM is 9.5598, for RWP it is 15.2287, and for RTT it is 326.3809333. The standard deviation calculated for them is follows: 1.386214936 for RWP, 99.10833546 for RTT, and 0.973908618 for SPMBMM.

Figures 6.6, 6.7, 6.8, 6.9, and 6.10 show the behavior of the different mobility model when the message generation interval varies.

It is observed from Fig. 6.6 the delivery probability is very high for SPMBMM as in correlation with RWP and RTT when the message generation interval increases in the network. The average delivery predictability for SPMBMM is 0.78458, for RWP it is 0.163936667, and for RTT it is 0.429933333. The standard deviation defined for RWP, RTT, and SPMBMM is 0.006880071, 0.045615506, and 0.053196594, respectively.

Figure 6.7 states that the RWP and SPMBMM need lesser number of hops to communicate the messages in the network. But on the other side, RTT needs very high number of hops in the environment. The average hop count calculated for

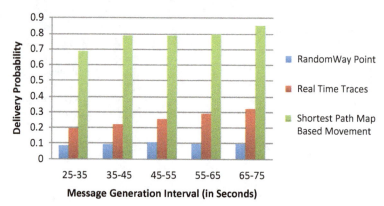

Fig. 6.6 Delivery probability versus message generation interval

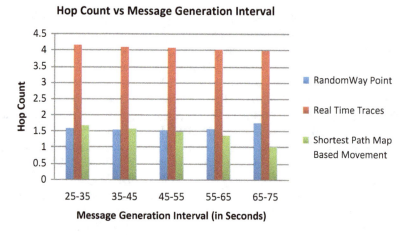

Fig. 6.7 Hop count versus message generation interval

RWP is 1.6073, for RTT it is 3.49924, and for SPMBMM is 1.6588. The standard deviation traced for the said mobility models when MGI varies is as 0.08133056 for RWP, 0.058574855 for RTT, and 0.227966843 for SPMBMM.

Figure 6.8 discusses about the message drop in the network with respect to MGI variation. The figure clearly shows that the RWP has minimum count of messages drop while performing routing from source to destination, whereas in RTT the messages drop are higher in the environment which results in the low delivery predictability under this scenario. On an average, 4519.4 is the message drop in RWP, 575965 for RTT, and 21031.4 for SPMBMM. 1221.976203 is the standard deviation for RWP, 176315.9688 for RTT, and 3914.860846 for SPMBMM.

6 Performance Analysis of A*-Based Hop Selection Technique in...

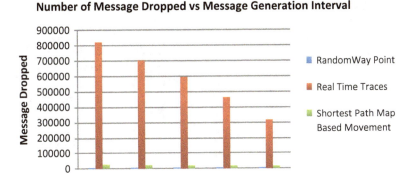

Fig. 6.8 Dropped messages versus message generation interval

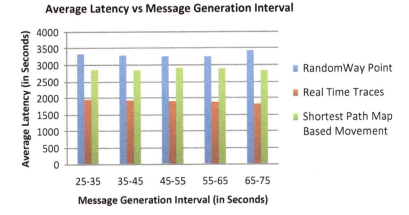

Fig. 6.9 Average latency versus message generation interval

Figure 6.9 discusses the behavior of message delay in A*OR protocol with the increase in MGI. From the graph, it is analyzed that the SPMBMM results in minimum delay while communicating the messages in the network. On an average, it is found that RTT results in 1893.60424 delay in the network, SPMBMM results in 2863.68338, and RWP results in 3309.04442. The standard deviation reviewed is as follows: 62.78098465 for RWP, 46.45699178 for RTT, and 28.40290486 for SPMBMM.

Figure 6.10 discusses about the overhead ratio with gradual increase in MGI. Graph reveals the minimum ratio of overhead in SPMBMM and maximum for RTT. The overhead ratio fabricated for SPMBMM is = 11.32478, for RWP = 19.54486667, and for RTT = 326.3809333 when the MGI varies. The standard deviation defined for them is as follows: for RWP it is 0.554479808, 52.80345451 for RTT, and 0.554479808 for SPMBMM.

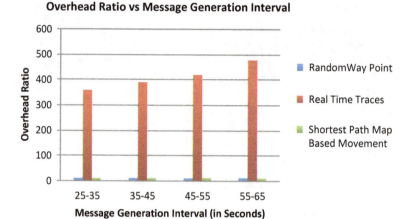

Fig. 6.10 Overhead ratio versus message generation interval

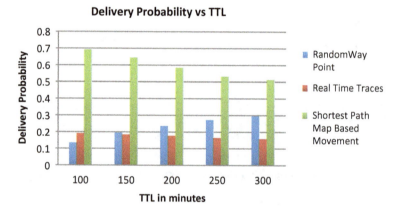

Fig. 6.11 Delivery probability versus TTL

Figures 6.11, 6.12, 6.13, 6.14, and 6.15 show the behavior of the different mobility model on the epidemic routing protocol when the TTL varies.

It is observed from Fig. 6.11 that the delivery probability is very high for SPMBMM as in comparison to RWP and RTT when the TTL varies in the network. The average delivery predictability for SPMBMM is 0.59224, for RWP it is 0.22632, and for RTT it is 0.17486. The standard deviation defined for RWP, RTT, and SPMBMM is 0.05727685, 0.012316103, and 0.067310047, respectively.

Figure 6.12 shows that RWP needs lower number of average hops for communicating the message in the network. However, it is still comparable for both RTT and SPMBMM at higher values of TTL. But in case of RTT, very high number of hops are required for transmitting the message in the network as the TTL increases gradually. The average hop count for RWP is 2.1317, RTT is 7.16866,

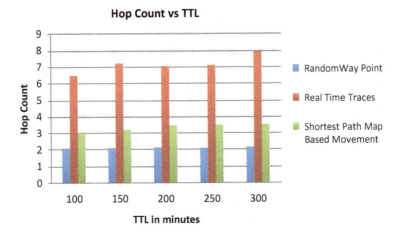

Fig. 6.12 Hop count versus TTL

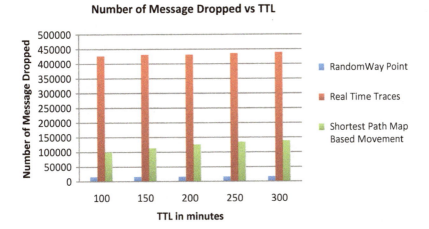

Fig. 6.13 Dropped messages versus TTL

and SPMBMM is 3.34838. The standard deviation traced when the TTL varies is as follows: 0.026854646 for RWP, 0.460785636 for RTT, and 0.203670581 for SPMBMM.

Figure 6.13 discusses about the drop rate of messages in the network with respect to TTL variation. The figure clearly shows that the RWP has dribbled/dropped minimum number of messages while conveying between source and destination. Whereas this drop is highest in RTT which results in low delivery predictability under this scenario. On an average, 16225.2 is the message drop in RWP, 436005 for RTT, and 122456.8 for SPMBMM. 407.1399759 is the standard deviation for RWP, 3243.3707 for RTT, and 14182.18253 for SPMBMM.

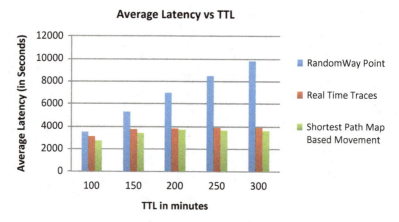

Fig. 6.14 Average latency versus TTL

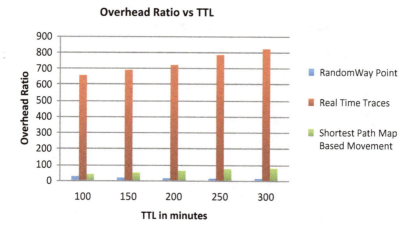

Fig. 6.15 Overhead ratio versus TTL

From Figure 6.14, variation in delay of messages is observed with respect to gradual increase in TTL. It is figured out from the graph that SPMBMM has minimum average delay as in comparison to rest of two RWP and RTT. The average latency is calculated for SPMBMM, RWP, and RTT, and found that SPMBMM has least average latency and RTT has second minimum average latency that is 3445.54574 and 3657.59008, respectively. RWP has maximum average delay time while communicating the message in the network that is 6805.92264. The standard deviation reviewed for different mobility models is as follows: 2231.457582 for RWP, 332.4044092 for RTT, and 283.66651 for SPMBMM.

Figure 6.15 discusses about the overhead ratio of the epidemic routing protocol in the network with respect to variation in the TTL. From the graph, it is analyzed that the RWP has minimum overhead ratio as compared to SPMBMM and RTT.

On an average, the overhead ratio calculated for RWP is 32.0676, for SPMBMM it is 61.84098, and for RTT it is 1223.754667. The standard deviation calculated for them is as follows: 4.880521917 for RWP, 59.93156237 for RTT, and 13.8281097 for SPMBMM.

6.5 Conclusion and Future Work

Mobility models play vital role in evaluating the performance of opportunistic network. Mobility models are characterized according to the real environment and they are categorized on the basis of the movement of nodes. In this chapter, author has discussed various mobility movement models (RWP, RTT, and SPMBMM) and their impact on the A*OR protocol is presented. Results show that the performance of A*OR is outstanding in terms of delivered messages, hop count, and overhead ratio under SPMBMM implementation. The average latency is incredibly low in RTT implementation as in comparison to SPMBMM and RWP. Overall, it is concluded that the SPMBMM mobility model produces optimum results for the A*OR protocol. As future work, inclination will be towards implying the security and energy aspects in this routing protocol.

References

1. C.-M. Huang, K.-C. Lan, C.-Z, Tsai, A Survey of opportunistic networks, in Proceedings of the 22nd International Conference on Advanced Information Networking and Applications – Workshops (IEEE, Piscataway, 2008), pp. 1672–1677
2. M. Demmer, J. Ott, S. Perreault, Delay tolerant networking TCP convergence layer protocol. Experimental RFC 7242, 2014
3. Y. Cao, Z. Sun, Routing in delay/disruption tolerant networks: a taxonomy, survey and challenges. IEEE Commun. Surv. Tutorials 15(2), 654–677 (2013)
4. J. Yick, B. Mukherjee, D. Ghosal, Wireless sensor network survey. Comput. Netw. 52(12), 2292–2330 (2008)
5. K. Fall, S. Farrell, DTN: an architectural retrospective. IEEE J. Sel. Areas Commun. 26(5), 828–836 (2008)
6. M. Musolesi, C. Mascolo, Designing mobility models based on social network theory. ACM SIGMOBILE Mobile Comput. Commun. Rev. 11(3), 59–70 (2007)
7. T. Camp, J. Boleng, V. Davies, A survey of mobility models for ad hoc network research. Wirel. Commun. Mob. Comput. 2(5), 483–502 (2002)
8. C. Song, Z. Qu, N. Blumm, A.Barabasi, Limits of predictability in human mobility. Science 327(5968), 1018–1021 (2010). http://dx.doi.org/10.1126/science.1177170
9. S. Batabyal, P. Bhaumik, Mobility models, traces and impact of mobility on opportunistic routing algorithms: a survey. IEEE Commun. Surv. Tutorials 17(3), 1679–1707 (2015). http://dx.doi.org/10.1109/COMST.2015.2419819
10. P. Kuchhal, S.K. Dhurandher, S.J. Borah, I. Woungang, S. Jain, S. Gupta, A* search based next hop selection for routing in opportunistic networks. Int. J. Space-Based Situat. Comput. 7(3), 177–186 (2017)

11. M. Kim, D. Kotz, S. Kim, Extracting a mobility model from real user traces, in Proceedings IEEE INFOCOM 2006. 25TH IEEE International Conference on Computer Communications (IEEE, Piscataway, 2006), pp. 1–13
12. A. Keranen, Keranen A. Opportunistic Network Environment Simulator. Special Assignment Report, Helsinki University of Technology, Department of Communications and Networking, 2008
13. A. Keranen, J. Ott, Increasing Reality for DTN Protocol Simulations. Special Technical Report, Helsinki University of Technology, Networking Laboratory, 2007

Chapter 7
Data Loss Prevention Using Document Semantic Signature

Hanan Alhindi, Issa Traore, and Isaac Woungang

Abbreviations

BM   Boyer Moore algorithm
CBSD   Component-based software development
CF   Concept vector file
CM   Concept map
CS   Cosine similarity
CT   Concept tree
DCT   Document concept tree
DL   Ontology description logics
DLP   Data loss prevention
DR   Detection rate
DSS   Document semantic signature
FDR   False discovery rate
FIBO   Financial Industry Business Ontology
FNR   False negative rate
FPR   False Positive rate
IDF   Inverse document frequency
IDS   Intrusion detection systems
KB   Knowledge base
KDE   Kernel density estimation

H. Alhindi (✉) · I. Traore
Electrical and Computer Engineering Department, University of Victoria, Victoria, BC, Canada
e-mail: halhindi@uvic.ca

I. Woungang
Department of Computer Science, Ryerson University, Toronto, ON, Canada

© Springer Nature Switzerland AG 2019
I. Woungang, S. K. Dhurandher (eds.), *2nd International Conference on Wireless Intelligent and Distributed Environment for Communication*, Lecture Notes on Data Engineering and Communications Technologies 27,
https://doi.org/10.1007/978-3-030-11437-4_7

NIDS	Network-based intrusion detection system
NTAC	National Threat Assessment Center
OWL	Ontology web language
RDF	Resource description framework
RNCVM	Relevancy nodes-based concept vector model
SIDD	Sensitive information dissemination detection
SVM	Support vector machines
SW	Smith–Waterman algorithm
TF	Term frequency
TF-IDF	Term frequency inverse document frequency

7.1 Introduction

7.1.1 Context

Malicious insiders are people with legitimate access to information and who then abuse their privileges to damage or steal an organization's resources and assets. Even though insider threat events are less widely publicized and are relatively more infrequent than external attacks, they usually are much more severe, causing a tremendous amount of damage to an institution or a business.

There are two categories of insider detection approaches: behavioral and technological [1]. Behavioral approaches use the psychological profiles of the perpetrators from job interviews and from monitoring the activity of the staff. Technological measures consist of deploying and operating monitoring devices or software to detect intrusions and policy violations. The main limitation of the existing technological approaches is that insider behavior does not necessarily look abnormal and hence may not be detectable by traditional security monitoring appliances.

Existing technological approaches rely on a wide spectrum of tools, ranging from traditional intrusion detection systems (IDS), security events logging and auditing, to data loss prevention (DLP) systems.

Our approach to protect against illicit data transfer falls under the DLP category. The DLP approach consists of monitoring data transfers by end users to ensure that sensitive information is not sent outside the organization network. Existing DLP systems provide insider detection and prevention capabilities by implementing organizational, business, and regulatory policies in the form of a series of predefined or customizable rules. The monitored data are matched against the rules, and a decision is made about if a data leak occurred or not.

There are two kinds of data matching: structured and unstructured data matching. Structured data follow clearly defined formats (e.g., SSN, credit card numbers, and telephone numbers), whereas unstructured data do not (e.g., e-mail content, source codes, and media files). Structured data matching is typically conducted using regular expressions derived from the underlying predefined format. In contrast, the lack of a predefined format makes unstructured data matching more challenging.

7 Data Loss Prevention Using Document Semantic Signature 77

The only option that is now used for unstructured data is to fingerprint the data by generating a cryptographic hash, which is stored in a secure location and then later compared against monitored data to identify a potential data leak.

7.1.2 Research Problem and Contribution

Existing DLP schemes monitor the file name or specific data formats or keywords contained in the file. However, these approaches fail to detect a data transfer when the original information is altered, rewritten, or reworded using synonyms or code words. For instance, such systems will fail to detect a situation where an insider reads classified information (that she or he is allowed to read) and transcribes a summary of this information in a new container (e.g., file, and e-mail) using a different lexicon or terminologies. We address the abovementioned shortcoming by proposing a DLP approach that relies on a new document content fingerprinting scheme called a document semantic signature (DSS). The DSS is derived by extracting and summarizing the semantic or meaning of the knowledge contained in a file or other container (e.g., e-mail, and repositories). The DSS is updated dynamically as the knowledge changes.

In contrast to the existing DLP schemes, our proposed approach allows for the tracking of malicious data transfers or exfiltration by monitoring the knowledge semantic rather than the container (i.e., physical or logical file) or selected keywords.

7.1.3 Approach Overview

Our content monitoring scheme uses domain ontologies to capture and encode the semantic of the knowledge or information being protected. In this context, an ontology is a formal representation of a set of concepts and the relations between these concepts in a domain of knowledge. We use domain-specific ontologies for the different kinds of knowledge being protected by a given organization (e.g., defense, healthcare, and finance). There are a few existing insider-related ontologies in the literature, such as the insider threat indicator ontology developed by CERT/CC [2]. However, these ontologies describe domains and concepts related to the creation and operation of insider-prevention tools, policies, and models. However, our interest lies in describing the actual knowledge or data that are being protected, and as such, we use ontologies specific to the corresponding knowledge domains.

The DSS will be derived by extracting a summarized representation of the semantics model for a given file or content. The DLP system will monitor newly generated contents (e.g., a new e-mail being written or new file creation) and track malicious data transfers between user accounts or data exfiltration between an insider's account and output channels (e.g., e-mails, printers, and online repositories) by comparing the DSS of the transferred data against the DSS of the sensitive documents.

The remaining sections of the current paper are organized as follows: Section 7.2 gives an overview of the literature and introduces related works on insider threat detection and DLP. Then, Sect. 7.3 presents and describes our proposed model for data loss prevention. Next, Sect. 7.4 describes the experimental evaluation of the proposed model and discusses its results. Finally, Sect. 7.5 provides a conclusion.

7.2 Related Works

In this section, we summarize and discuss the related work on insider detection, DLP, and ontological search.

7.2.1 On Insider Threat Detection and Prediction

Several insider threat prediction and detection models have been proposed. Most focus on analyzing user activity logs at the host level [3, 4] or at the network level [5].

Kandias et al. [3] proposed a hybrid insider threat prediction model that combines real-time usage profiling and psychological profiling along with user taxonomy. Real-time usage profiling involves monitoring users' behaviors by analyzing operating system calls and extracting a behavioral pattern for each user. Psychological profiling is based on the social learning theory and involves measuring the user's sophistication and predisposition through a questionnaire and the user's stress level, which is done using a psychometric test. The proposed prediction model uses the extracted information from real-time usage, psychological profiling, and the user taxonomy as inputs to decide and identify possible malicious behavior. The decision-making algorithm can predict and score suspicious insiders based on three factors: motive, opportunity, and capability.

Udoeyop [4] developed an insider detection approach by tracking threatening abnormal behavior through user activity monitoring. In the proposed approach, user activity information related to the hardware, process, network, and file system are collected. Various features are extracted from the data, such as processor usage, memory usage, hard drive usage, process threads, file system, network IP, and network port profiles. Normal behavioral profiles are constructed for each user using K-means clustering and kernel density estimation (KDE). In this approach, any new user behavior is collected and compared with the user's normal profile and flagged as abnormal if there is significant deviation.

Ragavan [6] introduced an insider threat mitigation model that monitors and prevents malicious write operations on sensitive data, here by using a log analysis and dependency graph. The proposed log model stores each write operation that is done on any data item into log files and then checks the number of changes on that data item according to the assigned threshold for each data item. The limitation of the proposed model is that it was tested on a synthetic dataset that could miss important characteristics of real-world datasets.

Liu et al. [5] developed a multilevel framework called the sensitive information dissemination detection (SIDD) system, which mainly aims at detecting and preventing sensitive data leaks in a protected network. The proposed detection system is placed at an egress point of the network to parse outgoing traffic and filter transferred sensitive content. Network traffic features are analyzed to detect the existence of covert channels. Some insider threat vectors use steganography techniques to hide the fact that the communication has occurred. In this case, the system uses steganalysis to detect these hidden channels.

Whereas the above approaches take a broader look at user activity (i.e., system calls, processes, and networks), we focus on leaks related to specific content, which is much narrower in scope but still important.

7.2.2 On DLP

Raman et al. [7] outlined the importance of the data leak prevention discipline by reviewing previous research, defining unsolved problems, introducing the challenges behind this problem, and motivating academic research to find a solution. Raman et al. [7] clarified that the goal behind data leak prevention is to detect and protect the resources, whereas the goal of intrusion detection is to detect the illegitimate users and protect the system from their activities.

Liu and Kuhn [8] discussed DLP challenges and defined several types of lost data, including leaked, disappeared, or damaged data. In addition, the authors identified three data loss modes, including data at rest, data at the endpoint, and data in motion, to find the best practices and solution capabilities to address the underlying problem.

Hart et al. [9] proposed a DLP approach that uses a machine-learning algorithm, specifically support vector machines (SVM), to learn and automatically classify both structured and unstructured sensitive information; the monitored information is classified as either public or private. Models are trained using an initial set of public and private documents, and the trained classifiers are later used to recognize sensitive (i.e., private) documents from nonsensitive ones. Hart et al. [9] introduced a new training technique, the supplement and adjust, which enables better discrimination between sensitive and nonsensitive data. The proposed approach was evaluated using five different datasets, yielding on average a false positive rate (FPR) of 0.46%, a false negative rate (FNR) of 1.6%, and a false discovery rate (FDR) of 0.47%. The FDR is defined as the ratio between the number of false positive and the sum of the numbers of false positive and true positive.

The authors claimed that they developed the first publicly available corpora for the evaluation of DLP systems. However, when contacted, the authors were only able to point to the Enron e-mail dataset, which was published elsewhere. They mentioned they could not share the private (or sensitive) subset of their evaluation corpora because of privacy and confidentiality reasons.

Some of the limitations of using machine learning for DLP are that for the model to be effective, enough representative samples of the sensitive data must be available to train the classifier, and the classifier may need to be retrained every time there is a significant change in the characteristics of the sensitive data. In contrast, our approach does not have this constraint because it depends only on the semantics of the data, regardless of its amount and future changes. Our model dynamically updates the semantic model as the content of the sensitive data evolves.

Stamati-Koromina et al. [10] proposed a data leak prevention model able to detect sensitive leaked data via e-mail messages using steganography. When the user sends an e-mail, the system scans, monitors, and logs the outgoing e-mail and its attachments. This model uses an SMTP proxy server, along with other online tools to obtain the attached e-mail's images and check if there is any embedded sensitive data inside these images. If the system detects a steganography payload in the attachment, the system will prevent sensitive data leaks by directly marking the e-mail as sensitive, sending an alert to the administrator, and terminating the transmission of the e-mail.

Canbay et al. [11] developed a DLP system in Turkish. In the proposed approach, the model is trained initially by generating a list of sensitive words from sensitive documents by computing and analyzing the term frequency and inverse document frequency (TF-IDF) metric. Detection is then carried out by comparing a monitored document against the trained model, aiming to locate any modifications of sensitive words, including by adding, deleting, and altering characters. The detection relies on using the Boyer Moore (BM) algorithm to search for explicit sensitive strings and the Smith–Waterman (SW) algorithm to detect altered sensitive strings. The proposed approach was evaluated on a dataset consisting of 180 documents covering different topics and yielded 100% recall and 98% accuracy. A key limitation of this approach is the inability to detect a modification when the semantic remains unchanged, for instance, by replacing a word with a synonym. In contrast, our proposed approach can detect this form of data leak because it relies on monitoring the semantic content.

7.2.3 On Ontology-Based Search and Information Retrieval

Our proposed approach uses ontology-based search capabilities to search for terms and their semantics, and there is an existing body of literature on ontology-based searches. These approaches cover information retrieval from a general perspective without a particular focus on DLP.

Vodithala and Pabboju [12] proposed an ontology-based search approach that relies on searching and retrieving information based on a keyword and its associated semantic keywords. The proposed approach was used in software engineering in the context of component-based software development (CBSD). CBSD aims at maximizing the reuse of software components to save time and reduce development cost. This work inspired our proposed model in providing a lexicon of keywords (i.e., a dictionary of concepts and their synonyms) when the system searches for a keyword in an ontology to retrieve related semantic keywords.

Fernández et al. [13] proposed an ontology-based information retrieval model that involves indexing, querying, searching, and ranking phases to enhance the semantic search in the web environment. The enhanced information retrieval model uses the domain knowledge base (KB) with an SPARQL query to obtain a set of tuples that are used to retrieve the documents that contain a keyword and its related semantic keywords from large document repositories. Our proposed model uses a similar retrieval technique, but we do not use SPARQL, which turns out to be difficult to configure and not adequate for our purpose. Instead, we use an ontology-based search approach that retrieves words and their related semantics by looking for these terms through two main ontology components: class and definition components.

7.2.4 On Ontology Management and Semantic Models

Doing-Harris et al. [14] developed an ontology management system called semi-automated ontology management (SEAM), which provides information extraction from clinical and biomedical documents based on OWL ontology files. SEAM uses natural language processing to extract terms and their relations. In this approach, a TF-IDF vector is generated for each N-gram, containing one entry for each document. An entry (i.e., TF-IDF value) in the vector for a given document is the occurrence frequency of the n-gram in the document divided by the average frequency across all documents.

In contrast, our proposed model extracts information from documents based on RDF ontology files, generating a collection of document vectors obtained by retrieving the depth of all terms with respect to a specific term in the ontology, where each vector corresponds to a unique concept from a document. The document vectors are combined into a matrix that represents the document's semantic signature.

Liu et al. [15] proposed a semantic model, the relevancy nodes-based concept vector model (RNCVM), where a concept vector is used to represent a particular concept node in a hierarchical structure. In this case, the concept vector of a specific node is based on the local density of all relevancy nodes in a taxonomy structure. In contrast, in our proposed model, we create the concept vector by defining the depth of all relevancy nodes of a specific node in the ontology. In our work and the paper by Liu et al. [15], the similarity between two concept nodes can be measured using their concept vectors. RNCVM with WordNet achieved a higher correlation value of 0.906 with human judgments when compared with several existing similarity measures. However, the model finds the similarity based on WordNet, which sometimes provides several synonyms that are not related to the domain of interest. Because of this, in our proposed model, we provide a lexicon for the specific domain of interest to facilitate retrieving the right synonyms of a concept.

7.3 Proposed DLP Model

7.3.1 Ontology Concept Tree

An ontology is a formal representation in a hierarchical structure of a set of concepts and the relations between these concepts in a specific knowledge domain [16]. The ontology allows for a representation of the concepts contained in a document. The relationships between these concepts provide the meaning of the content, also known as the semantic. Similar concepts or classes in the ontology are structured in a taxonomy structure called a concept tree. A concept tree (CT) describes the abstraction relationship (i.e., generalization/specialization) between similar concepts using a hierarchical structure. The root of the tree corresponds to the most abstract form of the concept, while intermediary nodes correspond to refined concepts, and leaf nodes correspond to instances.

An ontology may consist of several concept trees, each describing a group of related concepts. The collection of concept trees can be grouped into a larger tree representing the ontology. In this case, an abstract node serves as the root of the large ontology concept tree. This root node is commonly referred to as the Thing, which is more of a placeholder or abstraction that allows for the bundling of the different individual concept nodes in a single larger tree structure representative of the entire ontology.

Let $O = (C, R)$ denote the ontological tree for ontology O, where C and R correspond to the set of all concepts and the set of relationships among the concepts, respectively. By default, C contains at least the abstract Thing as the root of the tree: Thing $\in C$.

Given a concept node $c \in C$, let ancestors(c) and descendants(c) denote the set of all ancestor nodes and the set of all descendant nodes of c in the ontology tree, respectively. We define the relevancy nodes for c as follows:

$$relevancy(c) = ancestors(c) \cup descendants(c) \cup \{c\}$$

The relevancy nodes for a specific concept node include the root node, the ancestor nodes, the descendant nodes, and the node itself [15].

Let depth(c) denote the depth of the concept node $c \in C$ in the ontology, which represents the relative position of the node in the ontology concept tree with respect to the root, which is an integer value. The node's depth can be defined as the number of all nodes in the path from a specific concept node to the root (including the node itself and the root).

7.3.2 Document Concept Map

The ontology, through the collection of concept trees, provides a generic characterization of the knowledge that needs to be protected against leaks. The specific knowledge (actual files, databases, etc.) that needs to be protected is represented by its content semantics.

For a given document, we extract the concepts involved and represent its semantics using a semantic network, which is a structure used for representing knowledge as a pattern of interconnected nodes and arcs. We use a particular form of a semantic network, known as a concept map (CM), which is a directed graph where nodes represent concepts, and arcs represent relationships among them.

The steps to extract the set of concepts to derive the concept map for a document are as follows:

- Preprocess the document's content by removing metadata. For instance, for e-mails, this involves removing e-mail headers and keeping only the e-mail's body.
- Getting sentences consists of dividing the document content into separate sentences.
- Removing stop words filters out the sentences from the most common words in English based on a stop word list [17].
- Stemming reduces the number of words by deriving the roots of them.
- Creating the concept file is done after applying the aforementioned steps, saving the derived concepts in a text file.

7.3.3 Document Concept Tree

The document concept tree (DCT) captures the semantic of the document's content relative to a specific domain of knowledge, which is represented by an ontology. Given a document, the DCT is constructed by extracting all the concepts from the document concept map that are available in the ontology. As part of this process, synonymous concepts are replaced by matching the concepts available in the ontology [8].

Given an ontology $O = (C, R)$, the concept tree for a document d is defined as a triple $CT(d) = \{\{Thing\}, C_d, R_d\}$, where:

$C_d = \{c_1, c_2, \ldots, c_n\}$ is a set of concepts; each concept $c_i \in C_d$ is a word or phrase, and it is unique in C_d; also $C_d \subseteq C$.

$R_d = \{r_1, r_2, \ldots, r_t\}$ is a set of relationships among the concepts; each relationship $r_j \in R_d = (c_p, c_q, l_j)$, $p \neq q$, $1 \leq p, q \leq n$, $1 \leq j \leq t$, connects two concepts $c_p, c_q \in C$. Label l_j is a term that labels the relationship r_j.

Each document concept tree contains, by default, as its root, "Thing," the root of the ontology.

Algorithm 1 summarizes the steps for constructing the concept map for a given document, as shown below.

Algorithm 1 Summarizing document and extracting concept map file

---

```
/* arrList is an ArrayList of String */
/* file is a text File */
/* ConceptMap is a text file which has a set of extracted concepts from file*/
Input: void
Output: void

1:  procedure SUMMARY()
2:      File file ← loadFile(filepath);
3:      arrList ← setdocument(file, arrList);
4:      arrList ← GetSentences(file);
5:      arrList removestopwords(arr List);
6:      Tokenization();
7:      Stemming();
8:      Significants;
9:      File ConceptMap ← GetKeywordsToOutputFile();
10: end procedure=0
```

The runtime complexity of Algorithm 1 is $O(n\alpha^2)$, where n is the total number of concepts in a document, and α is the total number of sentences in a file. Its space complexity is $O(\alpha + n^2)$. The time and space complexity of Algorithm 1 are both quadratic functions.

7.3.4 Document Semantic Signature

Given a document d, the semantic signature of the document $SS(d)$ objectively captures the relevancy of each of the nodes c_i of the document concept tree with respect to each of the nodes c_j of the ontology concept tree. It is defined as the following matrix:

$$SS(d) = \begin{bmatrix} v_{ij} \end{bmatrix} \begin{matrix} 1 \leq i \leq n \\ 1 \leq j \leq m \end{matrix} \tag{7.1}$$

$$\text{where} \quad v_{ij} = \begin{cases} \text{depth}(c_i) & \text{if } c_j \in \text{relevancy}(c_i) \\ 0 & \text{otherwise} \end{cases} \tag{7.2}$$

and n and m correspond to the total numbers of concepts in the document and ontology concept trees, respectively. A row in the $SS(d)$ matrix is referred to as a concept vector, that is, row i ($1 \leq i \leq n$) corresponds to the concept vector for concept c_i.

7.3.5 Semantic Signature Matching

Given a set of documents $M = (d_1, \ldots, d_x)$ considered sensitive and that are being protected, we extract from each of the documents their semantic signature. Let n_{d_i} denote the number of concepts involved in the concept map of document d_i. So, each $SS(d_i)$ is an $n_{d_i} \times m$ matrix.

The set of semantic signatures represents the reference signature $SS(M) = (SS(d_1), \ldots, SS(d_x))$.

Note that the matrices corresponding to the semantic do not necessarily have the same number of rows because the number of concepts may be different in each document. In contrast, they all have the same number of columns m.

Given a suspected document d, the DLP consists of checking for similarity against the protected documents. This takes place by comparing the semantic signature $SS(d)$ against the reference signature $SS(M)$.

The matching consists of tracking the occurrence of each of the concept vectors of the monitored or suspicious document d for each of the semantic signatures in the reference signature.

Given i $(1 \leq i \leq n)$, concept vector $v_i = [v_{ij}]_{1 \leq j \leq m}$ from $SS(d)$ occurs in semantic signature $SS(d_k)$ from $SS(M)$ if one of the rows in $SS(d_k)$ exactly matches v_i. The matching involves initially calculating and aggregating the cosine similarity (CS) between the concept vectors of the documents as:

$$CS(SS(d), SS(d_k)) = \frac{\sum_{l=1}^{n} \sum_{r=1}^{n_{d_k}} CS(v_l(d), v_r(d_k))}{n \times n_{d_k}} \qquad (7.3)$$

where $CS(v_l(d), v_r(d_k))$ is the cosine similarity between the lth concept vector of $SS(d)$ and the rth concept vector of $SS(d_k)$, respectively, which is defined as follows:

$$CS(v_l(d), v_r(d_k)) = \frac{\sum_{j=1}^{m} v_{lj}(d) \times v_{rj}(d_k)}{\sqrt{\sum_{j=1}^{m} v_{lj}(d)^2} \times \sqrt{\sum_{j=1}^{m} v_{lj}(d_k)^2}} \qquad (7.4)$$

Then, let δ_{ik} denote the matching outcome, defined as follows:

$$\delta_{ik} = \begin{cases} 1 & \text{if} \quad CS(SS(d), SS(d_k)) = 1 \\ 0 & \text{otherwise} \end{cases} \qquad (7.5)$$

Next, we calculate the similarity of the monitored document with respect to the reference document by using two different metrics: a simple frequency model and the Jaccard index.

In the frequency model, the individual matching frequencies are determined and stored in a vector $F = [f_k]_{1 \leq k \leq x}$ where:

$$f_k = \frac{\sum_{i=1}^{n} \delta_{ik}}{n_{d_k}} \qquad (7.6)$$

In the Jaccard model, indices are calculated by comparing the monitored document signature with each of the reference document signatures, using the approach outlined above, whereby the number of matching concept vectors is tracked. The outcome of the comparisons is provided in a vector $J = [J_k]_{1 \leq k \leq x}$ where:

$$J_k = \frac{\sum_{i=1}^{n} \delta_{ik}}{n_{d_k} + n - \sum_{i=1}^{n} \delta_{ik}} \qquad (7.7)$$

Each of the similarity metrics are compared with separate predefined thresholds to establish similarity or dissimilarity and are then combined through the AND fusion model.

Let th_f and th_J denote the thresholds for the frequency and Jaccard metrics, respectively. The monitored document d is thought to contain a portion of some of the reference documents if the following exists:

$$\exists i \in \{1, \ldots, x\} \text{ such that } \left((J_k \geq th_J) \text{ and } \left(f_k \geq th_f \right) \right)$$

Algorithm 2 depicts the steps for extracting the concept tree of a document and measuring the similarity between a specific document and sensitive ones. The runtime complexity of Algorithm 2 is $O(m^3 x + q)$, whereas its space complexity is $O((n + x)m)$, where q is the total number of ontology files, m is the total number of concepts in the ontology, x is the total number of sensitive documents, and n is the total number of concepts in the document. The time complexity is cubic, while the space complexity is linear.

Algorithm 2 Extracting document concept tree and measuring similarity

---

    Input: void
    Output: void

1: **procedure** EXTRACTTREEMEASURESIMILARITY()
2:    *SynonymLoader();*
3:    *LoadRDFOntology();*
4:    *File TestFile ←LoadTestFile(TestFilePath);*
5:    *String FName ←TestFile.getName();*
6:    *Ontology.search(F Name);*
7:    *extractDocConceptTree(F Name);*
8:    *printV ectors(OntConceptVector Path, OntDocV ectorPath);*
9:    *calculateOntDocVecSize();*
10:   *printV ectors(ExtConceptV ector Path, ExtDocVectorPath);*
11:   *calculateExcDocVecSize();*
12:   *List < File > Ref Files ←LoadRef Dataset();*
13:   *int DocIndex←1;*
14:   **for** each File *Ref File: Ref Files* **do**
15:      *getDocVectorFromFile(Ref File.getCanonicalPath(), DocIndex);*
16:      *comparingConcepts(DocIndex);*
17:      *MeasureSimThreshold(DocIndex);*
18:      *MeasureConceptsSim(DocIndex);*

```
19:      MeasureJaccardSim(DocIndex);
20:      DocIndex++;
21:   end for
22:      printSimResults();
23: end procedure=0
```

7.4 Experiments

In this section, we present the evaluation procedures and dataset. Using the same dataset, we start by evaluating our proposed model and then compare it with two baseline models.

7.4.1 Dataset

To evaluate our model, we need a dataset that clearly identifies data that can be categorized as sensitive and leaked information. Unfortunately, there is no publicly available dataset that fully addresses this requirement. Hart et al. [9] used five datasets, including DynCorp, TM, Mormon, Enron e-mails, and Google private documents. Despite their claim that the datasets would be available publicly, after reaching out, they could not share the datasets, claiming privacy restrictions.

In our work, we used a real-life dataset, specifically a subset of the Enron e-mail dataset that focuses on business activities [18]. We grouped the e-mails by threads, with each thread consisting of an initial e-mail and a corresponding reply and forwarded messages. When a message is classified as sensitive, it is commonplace to categorize the messages belonging to the same thread as sensitive. For instance, the initial message can be categorized as confidential by the sender, and then, any other follow-up messages (part of the thread) would be treated in the same manner. Under this scenario, a DLP approach would consist of flagging the original messages as sensitive and then monitoring follow-up messages for possible leaks.

We took a similar view in structuring the aforementioned e-mail dataset to evaluate our proposed model. As mentioned above, we focused on only Enron business e-mails, which total approximately 447 e-mails. We grouped these e-mails into threads based on the initial e-mail, the responses, and the forwarded e-mails, yielding a total of 375 threads.

Figure 7.1 shows the total number of threads based on the number of e-mails. Each thread has either one, two, three, or four e-mails.

The goal of the evaluation is to test the ability of the proposed model to detect leaked information classified as sensitive while ignoring nonsensitive information. For a thread to be usable in testing a leak of sensitive information, we need at least two samples in it to use at least one as reference and the remaining for testing. Under

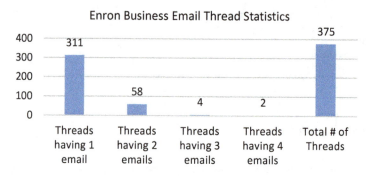

Fig. 7.1 Breakdown of number of threads with different number of e-mails

this constraint, we treated all the threads with a single message as nonsensitive, whereas the remaining threads were treated as sensitive.

7.4.2 Selected Ontology

Our approach depends on using an ontology that describes the domain of knowledge being protected. In accordance with the domain covered by our selected dataset, we chose the Financial Industry Business Ontology (FIBO) [19], which consists of 11 core domains, 49 modules, and 418 ontology files (FIBO). Because FIBO is a huge ontology, we have chosen the people, corporations, markets, and contracts modules as a partial representation of the ontology to implement in our model. A partial representation of the four modules of the FIBO ontology is shown in the appendix. Furthermore, we use a lexicon of business synonym keywords in our proposed model to assist in finding semantic keywords [20].

7.4.3 Evaluation Approach and Metrics

To assess the performance of our DLP approach, we calculate the DR and FPR. The DR measures the ability of the model to detect data leaks, while the FPR measures its ability to limit false alarms. A false positive occurs when a nonsensitive document is classified as sensitive. A false negative occurs when a sensitive document is falsely classified as nonsensitive; a true detection is just the opposite of a false negative, where a sensitive document is correctly classified as sensitive. The DR and FPR are calculated as follows:

$$FPR = \frac{\text{Number of non sensitive documents classified as sensitive}}{\text{Total number of nonsensitive documents}}$$

$$DR = \frac{\text{Number of sensitive documents classified as sensitive}}{\text{Total number of sensitive documents}}$$

We consider the initial e-mail from each thread that has two, three, and four e-mails as the sensitive reference e-mail dataset (A dataset) and the remaining e-mails in those threads as the sensitive testing e-mails dataset (B dataset). In addition, we consider the e-mails in the threads that have only one e-mail as the nonsensitive testing e-mail dataset (C dataset).

We conducted the evaluation by applying a twofold cross validation on our model, as follows. In the first round, we check the similarity of each e-mail in B against all e-mails in A (used as a reference) and calculate the number of e-mails in B that are flagged as dissimilar, as per our model (i.e., false negatives). Also, we check the similarity of each e-mail in C against all e-mails in A and calculate the number of e-mails in C that have similarity in A (i.e., false positives) as shown in Fig. 7.2.

In the second round, we flip between the A and B datasets. Then, we check the similarity of each e-mail in A against all e-mails in B and calculate the number of e-mails in A that do not have similarity in B (i.e., false negatives). Also, we check the similarity of each e-mail in C against all e-mails in B and calculate the number of e-mails in C that have similarity in B (i.e., false positives) as shown in Fig. 7.2.

7.4.4 Model Evaluation Results

In each round, we calculate the pair (FPR, DR) and then obtain the overall (FPR, DR) by calculating the average over the twofold cross validation rounds. In all experiments, we use the selected four modules of the FIBO ontology. We conducted different experiments by separately assessing the impact of the individual similarity

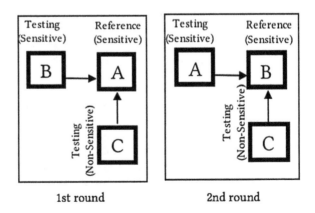

Fig. 7.2 Twofold cross validation

involved in our model and their combination. For the fusion of the similarity, we compare the AND model proposed in our work against the OR model.

Experiment 1 Tables 7.1 and 7.2 show the results of separately applying in our model the frequency and Jaccard similarity metrics, respectively, by varying the threshold values.

From these results, we found that applying our model based on only frequency similarity gives a higher DR and FPR than when using the Jaccard similarity only. In fact, the results of applying Jaccard only are satisfying, getting a high DR and low FPR.

Experiment 2 In this experiment, we apply the combination of the two semantic similarity metrics (i.e., frequency and Jaccard metrics), as described above. Table 7.3 shows the results of the OR combination between frequency and Jaccard, whereas Table 7.4 shows the results of the AND combination between frequency and Jaccard.

Table 7.1 Experiment 1: Applying frequency similarity metric only by varying the threshold

	thf = 55	thf = 60	thf = 65	thf = 70	thf = 75	thf = 80	thf = 85	thf = 90
	Average	Average	Average	Average	Average	Average	Average	Average
DR (%)	91.36	91.36	91.36	89	89	89	87.46	86.74
FPR (%)	65.1	59.06	58.22	50.67	44.97	39.1	37.75	35.39

Table 7.2 Experiment 1: Applying Jaccard similarity metric only by varying the threshold

	thJ = 55	thJ = 60	thJ = 65	thJ = 70	thJ = 75	thJ = 80	thJ = 85	thJ = 90
	Average	Average	Average	Average	Average	Average	Average	Average
DR (%)	87.56	84.47	81.38	80.65	74.28	68.74	63.48	53.77
FPR (%)	11.25	6.71	3.69	3.02	2.35	2.02	1.18	1.18

Table 7.3 Experiment 2: Applying OR combination of frequency with Jaccard for different thresholds

	thf = thJ = 55	thf = thJ = 60	thf = thJ = 65	thf = thJ = 70	thf = thJ = 75	thf = thJ = 80	thf = thJ = 85	thf = thJ = 90
	Average	Average	Average	Average	Average	Average	Average	Average
DR (%)	91.36	91.36	91.36	89	89	89	87.46	86.74
FPR (%)	65.1	59.06	58.22	50.67	44.97	39.1	37.75	35.39

Table 7.4 Experiment 2: Applying AND combination of frequency with Jaccard for different thresholds

	thf = thJ = 55	thf = thJ = 60	thf = thJ = 65	thf = thJ = 70	thf = thJ = 75	thf = thJ = 80	thf = thJ = 85	thf = thJ = 90
	Average	Average	Average	Average	Average	Average	Average	Average
DR (%)	87.56	84.47	81.38	79.93	73.56	65.74	58.31	50.72
FPR (%)	11.245	6.71	3.69	3.02	2.35	2.015	1.18	1.18

7 Data Loss Prevention Using Document Semantic Signature

From the results, we found that the OR combination model gives a higher DR and FPR than the AND combination. The results for the AND combination depicted in Table 7.4 are very encouraging in terms of their high DR and low FPR.

7.4.5 Comparison with Baseline Models

In this section, we compare our model separately against two different baseline models commonly used in information retrieval, that is, term frequency (TF), and term frequency and inverse document frequency (TF-IDF).

TF is the occurrence count of a term in a document divided by the number of concepts in the document. TF-IDF helps even out the effects of too many or few frequently occurring terms.

Consider a set of sensitive documents $M = (d_1, \ldots, d_x)$ that must be protected. The TF of the node (or concept) c_j from the ontology concept tree is defined as follows:

$$\text{tf}(c_j)_{d_i} = \frac{n(c_j)_{d_i}}{|d_i|} \tag{7.8}$$

where $|d_i|$ denotes the number of concepts involved in the document d_i, and $n(c_j)$ is the number of times concept c_j occurs in d_i. Using the same notation as above, m corresponds to the total numbers of concepts in the ontology concept tree, and $1 \leq j \leq m$.

The inverse document frequency (IDF) for c_j with respect to M is computed as follows:

$$\text{idf}(c_j)_M = 1 + \log\left(\frac{x}{\{d : M | c_j \in d\}}\right) \tag{7.9}$$

When concept c_j does not appear in any documents, then the denominator of Eq. (7.9) will be equal to zero, which in turn gives an infinite value for $\text{idf}(c_j)$. To avoid this, we assign -1 to $\text{idf}(c_j)$ if c_j does not appear in any document.

The TF-IDF for c_j with respect to M is computed as follows:

$$\text{tf} - \text{idf}(c_j)_{d_i, M} = \text{tf}(c_j)_{d_i} \times \text{idf}(c_j)_M \tag{7.10}$$

The TF matrix for document d_i is defined as follows:

$$\text{TF}(d_i) = \left[\text{tf}(c_j)_{d_i}\right]_{1 \leq j \leq m} \tag{7.11}$$

Similarly, the TF-IDF matrix for document d_i is defined as:

$$TF - IDF(d_i) = \left[tf - idf(c_j)_{d_i, M} \right]_{1 \le j \le m} \qquad (7.12)$$

Given a document d being checked for data leaks against reference M, the CS is applied against each of the sensitive documents $d_i \in M$. This gives for TF:

$$CS(d, d_i) = \frac{\sum_{j=1}^{m} tf(c_j)_d \cdot tf(c_j)_{d_i}}{\sqrt{\sum_{j=1}^{m} (tf(c_j)_d)^2} \times \sqrt{\sum_{j=1}^{m} (tf(c_j)_{d_i})^2}} \qquad (7.13)$$

whereas for TF-IDF:

$$CS(d, d_i) = \frac{\sum_{j=1}^{m} \left(tf - idf(c_j)_{d,M} \times tf - idf(c_j)_{d_i, M} \right)}{\sqrt{\sum_{j=1}^{m} \left(tf - idf(c_j)_{d,M} \right)^2} \times \sqrt{\sum_{j=1}^{m} \left(tf - idf(c_j)_{d_i, M} \right)^2}} \qquad (7.14)$$

In this experiment, we apply the TF and TF-IDF baseline models separately according to the different threshold values of the cosine similarity of the TF vectors (Thr_1) and TF-IDF vectors (Thr_2). The TF and TF-IDF vectors are created based on the FIBO ontology concepts.

In the TF baseline model, we calculate the cosine similarity between the TF vectors of the monitored document d against each sensitive documents $d_i \in M$. The cosine similarity value is compared against a threshold to make a decision regarding the similarity of the documents. In other words, if the cosine similarity value between the monitored document d and at least one sensitive document $CS(d, d_i)$ is above a certain threshold (Thr_1), then we will consider document d to be similar to d_i; otherwise, they are not similar.

We apply the same steps for the TF-IDF baseline model by taking into account the TF-IDF vectors and TF-IDF threshold (Thr_2).

Tables 7.5 and 7.6 show the results of applying the TF and TF-IDF baseline models, respectively.

Table 7.5 TF baseline model for different thresholds

	Thr1 = 55	Thr1 = 60	Thr1 = 65	Thr1 = 70	Thr1 = 75	Thr1 = 80	Thr1 = 85	Thr1 = 90
	Average	Average	Average	Average	Average	Average	Average	Average
DR (%)	76.02	75.20	69.84	63.47	50.39	41.23	32.87	25.97
FPR (%)	67.96	62.76	59.06	53.36	39.60	31.54	25.34	21.82

Table 7.6 TF-IDF baseline model for different thresholds

	Thr2 = 55	Thr2 = 60	Thr2 = 65	Thr2 = 70	Thr2 = 75	Thr2 = 80	Thr2 = 85	Thr2 = 90
	Average	Average	Average	Average	Average	Average	Average	Average
DR (%)	77.66	70.75	64.49	53.58	43.59	37.51	31.42	27.42
FPR (%)	64.77	55.71	50.68	42.96	37.75	32.72	27.52	21.98

From the results, the TF and TF-IDF baseline models give both a high DR and high FPR. This shows that our semantic signature model outperforms the above baseline models.

The configuration yielding the best performance is Experiment 2, when we applied our model based on AND combination of both frequency and Jaccard coefficient similarity metrics. As a result, the best results using 60 for both of threshold values was FPR = 6.71% and DR = 84.465%. These results are very encouraging. However, we intend to further explore more experimental configurations to improve on these results.

7.5 Conclusion

One of the main threats faced by any organization that maintains sensitive digital assets is the threat posed by malicious insiders. Because the sensitive information in any organization is vulnerable to being leaked, damaged, or lost, it is necessary to secure this information, along with assigning desired privileges to each employee. Some of these insiders, whose aim is to threaten the organization's security, leverage their privileges to leak sensitive data. Some can do this by altering the sensitive file's content but keeping the same meaning to remove any suspicion around using or transferring that file. Existing DLP mechanisms are inefficient when confronted with this kind of data alteration.

In the current paper, we attempted to address the aforementioned challenge by developing a new DLP approach to monitor transmitted data by checking its content semantically. From previous works, there is a clear limitation in searching and comparing contents semantically.

Our proposed model extracts a summarized form of the semantic of each document, which we refer to as the document's semantic signature. By comparing the similarity of the signature of a monitored document against reference signatures for sensitive documents, we can effectively detect potential data leaks. The basic components of our proposed approach have been defined, developed, and evaluated against an existing public dataset, yielding very encouraging performance results. Several practical issues identified above must still be addressed. One such issue is that our current signature model is relative to a specific domain ontology, which captures only knowledge related to the corresponding domain. However, it is not unusual that a sensitive document may carry important information related to multiple different domains, thereby to different ontologies. In our future work, we will expand our semantic signature model to cover a situation where multiple ontologies will be linked to the same signature, rather than having to maintain multiple signatures for the same document.

Our current experiments were conducted using only one dataset and one ontology. We also intend to run our model on a variety of datasets that cover different knowledge domains.

Appendix: Examples

In this section, we illustrate the DSS model through practical examples. Table 7.7 shows the concepts from the partial FIBO ontology graph in Fig. 7.3, along with their node labels and depths.

Table 7.7 Partial FIBO ontology concepts, label, and depth

Label	Depth	Concept	Label	Depth	Concept
C_1	1	Thing	C_{36}	3	Religious Corporation
C_2	2	Country	C_{37}	3	Contract
C_3	2	Date	C_{38}	3	Market
C_4	2	Text	C_{39}	3	Contract Document
C_5	2	Commitment	C_{40}	3	Identity Document
C_6	2	Transferable Contract	C_{41}	3	Joint Stock Company
C_7	2	Formal Organization	C_{42}	3	Privately Held Company
C_8	2	Physical Location	C_{43}	3	Publicly Held Company
C_9	2	Facility	C_{44}	3	National Identification Number
C_{10}	2	Party In Role	C_{45}	3	Market Identifier
C_{11}	2	Autonomous Agent	C_{46}	3	Board Agreement
C_{12}	2	Physical Address	C_{47}	3	Registration Identifier
C_{13}	2	Legally Capable Person	C_{48}	4	Regulated Market
C_{14}	2	Contractual Element	C_{49}	4	Multilateral Trading Facility
C_{15}	2	Yes Or No	C_{50}	4	Organized Trading Facility
C_{16}	2	Mutual Agreement	C_{51}	4	Contract Principal
C_{17}	2	Monetary Amount	C_{52}	4	Contract Counterparty
C_{18}	2	Not For Profit Corporation	C_{53}	4	Adult
C_{19}	2	Agreement	C_{54}	4	Minor
C_{20}	2	Venue	C_{55}	4	Non-Binding Terms Set
C_{21}	2	Legal Document	C_{56}	4	Non-Binding Terms
C_{22}	2	Stock Corporation	C_{57}	4	Written Contract
C_{23}	2	Identifier	C_{58}	4	Mutual Contractual Agreement
C_{24}	2	XML Schema #string	C_{59}	4	Unilateral Contract
C_{25}	2	Organization Coverg Agreement	C_{60}	4	Verbal Contract
C_{26}	2	Jurisdiction	C_{61}	4	Birth Certificate
C_{27}	2	Organization Identifier	C_{62}	4	Drivers' License
C_{28}	3	Exchange	C_{63}	4	Passport
C_{29}	3	Contract Third Party	C_{64}	5	Registered Multilateral Trading Facility
C_{30}	3	Contract Party	C_{65}	5	Contract Originator
C_{31}	3	Person	C_{66}	5	Transferable Contract Holder
C_{32}	3	Conditions Precedent	C_{67}	5	Incapacitated Adult
C_{33}	3	Contract Terms Set	C_{68}	5	Legally Capable Adult
C_{34}	3	Contractual Commitment	C_{69}	5	Emancipated Minor
C_{35}	3	Contractual Definition	C_{70}	5	Promissory Note

7 Data Loss Prevention Using Document Semantic Signature

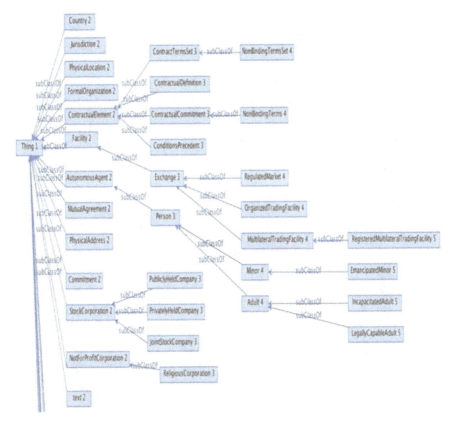

Fig. 7.3 Partial representation of the FIBO ontology tree

Figure 7.4 shows a sample document, specifically an e-mail from the Enron e-mail dataset; [18, 21] we refer to it as the reference (or sensitive) document d_1

The extracted concept file, document concept tree, and document semantic signature from the above e-mail sample are shown in Figs. 7.5 and 7.6.

To illustrate the matching process, assume that we have three sensitive documents in the reference (including the e-mail sample given above, d_1): $M = (d_1, d_2, d_3)$.

Using the above approach, we can generate the reference signature $SS(M) = \{SS(d_1), SS(d_2), SS(d_3)\}$.

Assume that we have two monitored documents, CF_1 and CF_2, that need to be matched against the reference signature. Figure 7.7 shows the matching process of the monitored document CF_1 against the reference signature $SS(M)$.

This figure shows the comparison process of each concept vector in the monitored document CF_1 against all sensitive documents' semantic signatures. If there is a match between the monitored concept vector and document's semantic signature, then the frequency will be incremented by one and saved to the frequency matrix and so on. Then, the same steps will be repeated for all concept vectors in the

```
Message-ID: <7879273.1075840857246.JavaMail.evans@thyme>
Date: Tue, 22 May 2001 19:37:00 -0700 (PDT)
From: edhearst@earthlink.net
To: louise.kitchen@enron.com
Subject:
Mime-Version: 1.0
Content-Type: text/plain; charset=us-ascii
Content-Transfer-Encoding: 7bit
X-From: Edward Hearst  <edhearst@earthlink.net>
X-To: louise.kitchen <louise.kitchen@enron.com>
X-cc:
X-bcc:
X-Folder: \ExMerge - Kitchen, Louise\'Americas\HR
X-Origin: KITCHEN-L
X-FileName: louise kitchen 2-7-02.pst

Dear Ms. Kitchen,

I am currently working as a V.P. at Commerce One managing the Global Trading
Web.  I have followed Enron's e-commerce activities for some time.  I am
currently exploring other career opportunities and thought I might be able
to contribute to Enron's online initiative.  I previously worked at the
House Energy and Commerce Committee, the FCC, the State Dept, and Jones, Day
Reavis & Pogue.  I not only have experience in the B2B world through my work
at Commerce One, but also in e-commerce generally, trade and
telecommunications through these other positions.  This combination of
experience could be helpful to Enron in expanding your B2B effort, including
governance and interoperability issues, developing new business, and in
dealing with domestic and international regulatory matters related to
e-commerce.

A copy of my resume is attached.  If you have a few moments to talk, I would
greatly appreciate it.

Thanks and best regards,
```

Fig. 7.4 Sample document from the Enron e-mail dataset

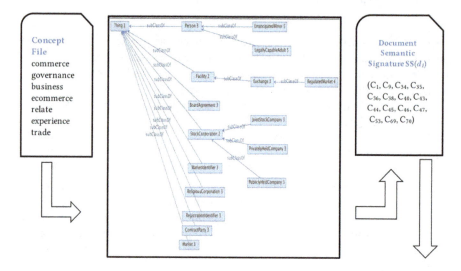

Fig. 7.5 Extracting concept tree and generating semantic signature for Reference Document $d1$

7 Data Loss Prevention Using Document Semantic Signature

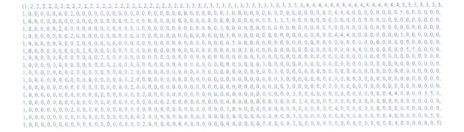

Fig. 7.6 Document semantic signature SS($d1$)

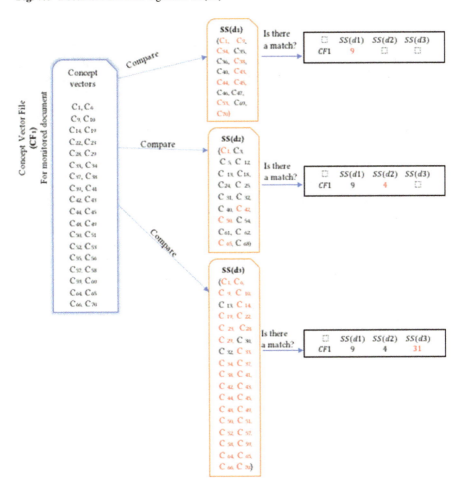

Fig. 7.7 Matching process of monitored document CF$_1$ against the reference signature

monitored document against the remaining documents' semantic signatures. As an example, CF_1 has nine matched concepts in sensitive document $SS(d1)$, four matched concepts in $SS(d2)$, and 31 matched concepts in $SS(d3)$.

The two matrices below represent the matching frequencies for the two monitored documents CF_1 and CF_2.

$$
\begin{array}{cccc}
 & SS\,(d_1) & SS\,(d_2) & SS\,(d_3) \\
CF_1 & 9 & 4 & 31
\end{array}
$$

$$
\begin{array}{cccc}
 & SS\,(d_1) & SS\,(d_2) & SS\,(d_3) \\
CF_2 & 3 & 2 & 6
\end{array}
$$

Also, the two matrices below show the frequency percentage for the monitored concept vector files CF_1 and CF_2, which show that the highest percentage of frequency of CF_1 is 91.18% in $SS(d_3)$, while the lowest frequency percentage is 22.22% in $SS(d_2)$. For the second monitored file CF_2, the highest frequency percentage is 20% in $SS(d_1)$, while the lowest is 11.11% in $SS(d_2)$.

$$
\begin{array}{cccc}
 & SS\,(d_1) & SS\,(d_2) & SS\,(d_3) \\
F\,(CF_1) & 60\% & 22.22\% & 91.18\%
\end{array}
$$

$$
\begin{array}{cccc}
 & SS\,(d_1) & SS\,(d_2) & SS\,(d_3) \\
F\,(CF_2) & 20\% & 11.11\% & 17.65\%
\end{array}
$$

In addition, the Jaccard index is calculated below for both monitored documents. The two matrices below show that the highest Jaccard index for CF_1 is 79.49% in $SS(d3)$, while the highest Jaccard index for CF_2 is 42.86% in $SS(d1)$.

$$
\begin{array}{cccc}
 & SS\,(d_1) & SS\,(d_2) & SS\,(d_3) \\
J\,(CF_1) & 21.43\% & 8\% & 79.49\%
\end{array}
$$

$$
\begin{array}{cccc}
 & SS\,(d_1) & SS\,(d_2) & SS\,(d_3) \\
J\,(CF_2) & 42.86\% & 8\% & 16.22\%
\end{array}
$$

From the measures above, our model will classify CF_1 as a suspicious file because the Frequency $F(CF_1) = 91.18\%$ against $SS(d3)$, which is higher than the threshold value 60%, and the Jaccard index $J(CF_1) = 79.49\%$ against $SS(d3)$, which is higher than the threshold value 60%, too.

References

1. E. Kowalski, D. Cappelli, A. Moore, *U.S. Secret Service and CERT/SEI Insider Threat Study: Illicit Cyber Activity in the Information Technology and Telecommunications Sector* (Carnegie Mellon Software Engineering Institute, Pittsburgh, 2008)
2. D.L. Costa, M.L. Collins, S.J. Perl, et al., *An Ontology for Insider Threat Indicators Development and Applications* (Carnegie-Mellon University, Pittsburgh, Software Engineering Inst, 2014)
3. M. Kandias, A. Mylonas, N. Virvilis, et al., An insider threat prediction model, in *International Conference on Trust, Privacy and Security in Digital Business*, (Springer, Cham, 2010), pp. 26–37
4. A.W. Udoeyop, Cyber Profiling for Insider Threat Detection. Master's Thesis, University of Tennessee (2010)
5. Y. Liu, C. Corbett, K. Chiang, et al., SIDD: A framework for detecting sensitive data exfiltration by an insider attack. System Sciences, 2009. HICSS'09. 42nd Hawaii International Conference on IEEE 2009, pp. 1–10
6. H. Ragavan, *Insider threat mitigation models based on thresholds and dependencies* (University of Arkansas, Fayetteville, 2012)
7. P. Raman, H.G. Kayacık, A. Somayaji, Understanding data leak prevention, in *6th Annual Symposium on Information Assurance (ASIA'11)* (2011), pp. 27–3
8. S. Liu, R. Kuhn, Data loss prevention. IT Professional **12**(2), 10–13 (2010)
9. M. Hart, P. Manadhata, R. Johnson, Text classification for data loss prevention, ed. by S. Fischer-Hübner, N. Hopper. PETS 2011. LNCS, vol. 6794 (2011), p 18–37
10. V. Stamati-Koromina, C. Ilioudis, R. Overill, et al., Insider threats in corporate environments: a case study for data leakage prevention, in *Proceedings of the Fifth Balkan Conference in Informatics*, (ACM, New York, 2012), pp. 271–274
11. Y. Canbay, H. Yazici, S. Sagiroglu, A Turkish language based data leakage prevention system. in *Digital Forensic and Security (ISDFS), 5th International Symposium* (IEEE, April 2017), pp. 1–6
12. S. Vodithala, S. Pabboju, A keyword ontology for retrieval of software components. Int. J. Control Theory Appl. **10**(19), 177–182 (2017)
13. M. Fernández, I. Cantador, V. López, et al., Semantically enhanced information retrieval: an ontology-based approach. Web Semant. Sci. Serv. Agents World Wide Web **9**(4), 434–452 (2011)
14. K. Doing-Harris, Y. Livnat, S. Meystre, Automated concept and relationship extraction for the semi-automated ontology management (SEAM) system. J. Biomed. Semant. **6**, 15 (2015)
15. H.Z. Liu, H. Bao, D. Xu, Concept vector for similarity measurement based on hierarchical domain structure. Comput. Inform. **30**(5), 881–900 (2012)
16. C. Corley, R. Mihalcea, Measuring the semantic similarity of texts. in *Proceedings of the ACL Workshop on Empirical Modeling of Semantic Equivalence and Entailment, Association for Computational Linguistics*, 2003, p 13–18
17. Onix, Onix Text Retrieval Toolkit API Reference (2017), http://www.lextek.com/manuals/onix/stopwords1.html, Accessed 14 Nov 2017
18. B. Klimt, Y. Yang, The Enron Corpus: a new dataset for email classification research, in *Machine learning, ECML 2004*, (Springer, Berlin, 2004), pp. 217–226
19. FIBO, Financial Industry Business Ontology (2017), https://www.edmcouncil.org/financialbusiness. Accessed 20 Oct 2017
20. Business Balls (2017), http://www.businessballs.com/business-thesaurus.htm. Accessed 19 Oct 2017
21. Enron Email Dataset (2017), http://www-2.cs.cmu.edu/~enron/. Accessed 20 Oct 2017

Chapter 8
Understanding Optimizations and Measuring Performances of PBKDF2

Andrea Francesco Iuorio and Andrea Visconti

8.1 Introduction

Although user-chosen passwords are often too short and lack enough entropy [1], they are still widely used for authentication purposes, thus making them vulnerable to brute-force or dictionary attacks. A possible solution to this issue is to adopt a key derivation function (KDF) which inputs a key material and generates a secure key [2]. In particular, [3] provides recommendations for the implementation of PBKDF2, a KDF which inputs a user-chosen password.

Even though in 2015 the *Password Hashing Competition* [4] selected a number of hashing schemes—e.g., Argon2 [5] (the winner of the competition), Catena [6], Lyra2 [7], yescrypt [8], and Makwa [9]—currently, PBKDF2 [3] still remains the most widely implemented and used in practice. For example, it is used in WiFi protected access [10], iOS passcodes [11], LUKS [12], and many others. In addition, KDF has been applied by researchers in mobile ad hoc network security not for detecting misbehaving nodes [13] but for securing the zone routing protocol [14]. In addition, in the internet of things (IoT) era users want to be able to access their accounts on all their devices, thus adopting password managers to remember and secure user-chosen passwords. Notice that several password manager applications [15–19] are based on PBKDF2 and a number of security and privacy concerns have to be addressed [20–22].

For slowing attackers down, PBKDF2 uses a random salt and an iteration count. The latter specifies the number of times a pseudorandom function (PRF) is iterated to generate a key of appropriate size. The iteration count is one of the most important

---

A. F. Iuorio · A. Visconti (✉)
Department of Computer Science, Università degli Studi di Milano, Milano, Italy
e-mail: andreafrancesco.iuorio@studenti.unimi.it; andrea.visconti@unimi.it

© Springer Nature Switzerland AG 2019
I. Woungang, S. K. Dhurandher (eds.), *2nd International Conference on Wireless Intelligent and Distributed Environment for Communication*, Lecture Notes on Data Engineering and Communications Technologies 27,
https://doi.org/10.1007/978-3-030-11437-4_8

parameters of PBKDF2. The choice of a high value slows attackers down but may negatively affect usability. In [23], NIST recommends a minimum of 1000 iterations for general purpose applications but suggests to select the iteration count value as large as possible. Interestingly, many applications define such a value a priori—for example, WPA2 sets the iteration count value to 4096 [10]—while others do not—e.g., the iteration count associated with iOS passcodes is calibrated to take about 80 ms [11].

In this paper, we focus on PBKDF2-HMAC-SHA-1, presenting the state-of-the-art research results achieved in the last 5 years. In particular, we introduce all cryptographic algorithms involved in the key derivation process. Then, we describe the optimization techniques used to speed up PBKDF2, HMAC, and SHA-1 in a GPU/CPU context. Finally, in order to measure the contributions provided by these optimizations, we develop an implementation of PBKDF2-HMAC-SHA-1 from scratch, execute a testing activity on consumer-grade hardware, and present the experimental results found.

The paper is organized as follows. In Sect. 8.2 we introduce the cryptographic algorithms involved in the key derivation process. In Sect. 8.3, we describe several optimizations published in literature that can be used to speed up a PBKDF2 implementation. In Sect. 8.4, we present our testing activities, describing both CPU and GPU implementations and showing the experimental result found. Finally, conclusions are drown in Sect. 8.5.

8.2   Cryptographic Preliminaries

8.2.1   PBKDF2

PBKDF2 is a password-based key derivation function: starting from a password, the algorithm generates a key of fixed length. PBKDF2 can be described as a chain of several instances of a pseudorandom function. In this paper we focus on PBKDF2-HMAC-SHA-1. Although in 2017 first practical technique for generating a collision of SHA-1 has been presented, HMAC-SHA-1 is still considered secure.

PBKDF2 is a password-based key derivation function described in PKCS #5 [3, 23]. For providing better resistance against brute-force attacks, PBKDF2 introduces CPU-intensive operations. These operations are based on an iterated pseudorandom function (PRF) which maps input values to a derived key. The most important properties to assure is that the iterated pseudorandom function is cycle free. If this is not so, a malicious user can avoid the CPU-intensive operations and, as described in [24, 25], get the derived key by executing a set of functionally equivalent instructions.

PBKDF2 inputs a pseudorandom function PRF, the user password p, a random salt s, an iteration count c, and the desired length len of the derived key. It outputs a derived key $DerKey$.

8 Understanding Optimizations and Measuring Performances of PBKDF2

$$Der\,Key = \text{PBKDF2}(PRF, p, s, c, len) \tag{8.1}$$

More precisely, the derived key is computed as follows:

$$Der\,Key = T_1||T_2||\ldots||T_{len}, \tag{8.2}$$

where

$$T_1 = Function(p, s, c, 1),$$

$$T_2 = Function(p, s, c, 2),$$

$$\vdots$$

$$T_{len} = Function(p, s, c, len).$$

Each single block T_i—i.e., $T_i = Function(p, s, c, i)$—is computed as

$$T_i = U_1 \oplus U_2 \oplus \ldots \oplus U_c, \tag{8.3}$$

where

$$U_1 = \text{PRF}(p, s||i),$$

$$U_2 = \text{PRF}(p, U_1),$$

$$\vdots$$

$$U_c = \text{PRF}(p, U_{c-1}).$$

The PRF adopted can be a hash function [26], cipher, or HMAC [27, 28], and [29]. In the sequel, we will refer to HMAC as PRF.

8.2.2 HMAC

An hash-based message authentication code (HMAC) is an algorithm for computing a message authentication code based on any iterated cryptographic hash function. The definition of HMAC [29] requires

- H: a cryptographic hash function;
- K: the secret key;
- $text$: the message to be authenticated.

As described in RFC 2104 [29], an HMAC can be defined as follows:

$$\text{HMAC} = H(K \oplus opad, H(K \oplus ipad, text)) \quad (8.4)$$

where H is the chosen hash function, K is the secret key, and *ipad*, *opad* are constant values—respectively, the byte 0x36 and 0x5C repeated 64 times. Recall that Eq. (8.4) can be expanded in the form:

$$h = H(K \oplus ipad \parallel text)$$

$$\text{HMAC} = H(K \oplus opad \parallel h)$$

In our performance tests, the hash function adopted will be SHA-1, thus making HMAC-SHA-1 the default pseudorandom function.

8.2.3 SHA-1

SHA-1 is a cryptographic hash function that inputs an arbitrarily long message M and outputs a 160-bit digest H. In order to provide the message digest, SHA-1 operates eighty times on five 32-bit words A, B, C, D, and E as shown in Fig. 8.1. Notice that F_t is defined by

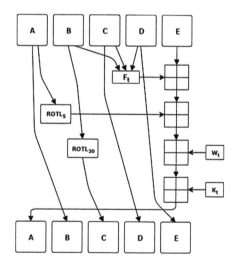

Fig. 8.1 A graphical representation of the SHA-1 algorithm

8 Understanding Optimizations and Measuring Performances of PBKDF2

$$\begin{cases} F_0 = (B \wedge C) \vee ((\neg B) \wedge D) & t \in [0 \ldots 19] \\ F_1 = (B \oplus C \oplus D) & t \in [20 \ldots 39] \\ F_2 = (B \wedge C) \vee (B \wedge D) \vee (C \wedge D) & t \in [40 \ldots 59] \\ F_3 = (B \oplus C \oplus D) & t \in [60 \ldots 79] \end{cases}$$

and K_t assume four constants value.[1]

Message M is processed in blocks of the size of 512 bits, namely sixteen 32-bit words $W_0, \ldots, W_{15}$, eventually padding the last block. More precisely, the last block is padded with one bit **1** first then, zero or more bits **0** so that its length is congruent to 448, modulo 512. The remaining 64 bits of the last 512-bit block represent the message length L. The SHA-1 algorithm expands 32-bit words $W_0, \ldots, W_{15}$ into eighty words using the follow message scheduling function:

$$W_i = ROTL^1(W_{i-3} \oplus W_{i-8} \oplus W_{i-14} \oplus W_{i-16}) \qquad i \in [16 \ldots 79] \qquad (8.5)$$

where $ROTL(x, n)$ is the left rotation of x by n bits. Notice that Eq. (8.5) requires to store eighty 32-bit words. If memory is limited (e.g., embedded devices and GPUs), an alternative method should be adopted. NIST suggests to regard $W_0, \ldots, W_{15}$ as a circular queue [26] and substitute Eq. (8.5) with the following:

$$\begin{cases} s = i \wedge MASK & i \in [16 \ldots 79] \\ W_s = ROTL^1(W_s \oplus W_{(s+2) \wedge MASK} \oplus W_{(s+8) \wedge MASK} \oplus W_{(s+13) \wedge MASK}) \end{cases}$$
$$(8.6)$$

where $MASK$ is set to the value $0x0F$ in Hex. Equation (8.6) requires only sixteen words, thus saving sixty-four 32-bit words of storage.

Further improvements have been presented in [25]. In particular, the authors suggest to replace Eq. (8.5) with

$$W[i] = \begin{cases} ROTL^1(W_{i-3} \oplus W_{i-8} \oplus W_{i-14} \oplus W_{i-16}) & i \in [16 \ldots 31] \\ ROTL^2(W_{i-6} \oplus W_{i-16} \oplus W_{i-28} \oplus W_{i-32}) & i \in [32 \ldots 63] \\ ROTL^4(W_{i-12} \oplus W_{i-32} \oplus W_{i-56} \oplus W_{i-64}) & i \in [64 \ldots 79] \end{cases} \qquad (8.7)$$

and then replace $W_{29}, W_{30}, W_{31}, W_{60}$, and W_{62} with the following and less expensive (we are reducing the number of XORs) equations:

$$\begin{cases} W_{29} = ROTL^2(W_{23}) \oplus k[29] \\ W_{30} = ROTL^2(W_{24} \oplus k[16]) \\ W_{31} = ROTL^2(W_{25} \oplus k[17]) \oplus k[31] \\ W_{60} = ROTL^4(W_{48} \oplus W_{28} \oplus W_0) \\ W_{62} = ROTL^4(W_{50} \oplus W_{30} \oplus W_0) \end{cases} \qquad (8.8)$$

---

[1] In this section, we partially describe the SHA-1 algorithm. Further details can be found in [26].

where $k[29] = ROTL^2(W_5) \oplus ROTL^1(W_{15})$, $k[16] = W_0 \oplus W_2$ (previously computed in W_{16}), $k[17] = W_1 \oplus W_3$ (previously computed in W_{17}), and finally $k[31] = ROTL^1(W_{15}) \oplus ROTL^2(W_{15})$.

In addition, [25] states that Eq. (8.6) can be replaced with the unfolded version:

$$
\begin{cases}
W_{16} = W_0^1 \oplus W_2^1 \oplus W_8^1 \oplus W_{13}^1 \\
W_{17} = W_1^1 \oplus W_3^1 \oplus W_9^1 \oplus W_{14}^1 \\
W_{18} = W_2^1 \oplus W_4^1 \oplus W_{10}^1 \oplus W_{15}^1 \\
W_{19} = W_0^2 \oplus W_2^2 \oplus W_3^1 \oplus W_5^1 \oplus W_8^2 \oplus W_{11}^1 \oplus W_{13}^2 \\
W_{20} = W_1^2 \oplus W_3^2 \oplus W_4^1 \oplus W_6^1 \oplus W_9^2 \oplus W_{12}^1 \oplus W_{14}^2 \\
W_{21} = W_2^2 \oplus W_4^2 \oplus W_5^1 \oplus W_7^1 \oplus W_{10}^2 \oplus W_{13}^1 \oplus W_{15}^2 \\
W_{22} = W_0^3 \oplus W_2^3 \oplus W_3^2 \oplus W_5^2 \oplus \cdots \oplus W_{11}^2 \oplus W_{13}^3 \oplus W_{14}^1 \\
W_{23} = W_1^3 \oplus W_3^3 \oplus W_4^2 \oplus W_6^2 \oplus \cdots \oplus W_{12}^2 \oplus W_{14}^3 \oplus W_{15}^1 \\
\cdots \\
W_{79} = W_0^8 \oplus W_0^{22} \oplus W_1^7 \oplus \cdots \oplus W_{15}^{14} \oplus W_{15}^{17} \oplus W_{15}^{18}
\end{cases}
\tag{8.9}
$$

where $W_i^j = ROTL^j(W_i)$. Notice that, although Eq. (8.9) increases the total number of XOR operations, if we compute PBKDF2-HMAC-SHA-1, it requires to store only five 32-bit words, namely $W_0, \dots, W_4$, because $W_6, \dots, W_{14}$ are equal to zero, and W_5, W_{15} are constant value. Therefore, this approach might be exploited by GPGPU programming.

8.3 Understanding Optimizations

PBKDF2 applies a pseudorandom function to generate cryptographically secure keys. Since in this process different cryptographic algorithms are involved (see Sect. 8.2), the optimization of one of these algorithms usually leads to interesting performance improvements in the key derivation process. But this is not always true. Indeed, some optimizations described in this section affect SHA-1 or HMAC-SHA-1 but have no effect on PBKDF2-HMAC-SHA-1. A crucial role is played by the context in which the code will be run, namely a GPU or CPU context. In fact, a specific algorithmic optimization may have no impact on GPU performances, while it has on CPU ones. Interestingly, however, the opposite is true as well.

Focusing on the state of the art of PBKDF2, HMAC, and SHA-1, in this section we briefly present the optimizations resulting to significant improvements.

8.3.1 PBKDF2 Optimizations

[OPT–01] Early Exit The execution time spent for computing a derived key does not only depend on the iteration count values. Indeed, also the number of fingerprints T_i required to compute a single iteration affects the total execution time. Assuming that we require a 256-bit derived key, two SHA-1 fingerprints are necessary— i.e., $DerKey = T_1 || T_2$, with T_1 and T_2 160-bit length each. Since blocks T_i are independent of each other, firstly we generate a block T_1 and then we compute the second if and only if T_1 is equal to the first part of the 256-bit derived key. If not so, the chosen password p is certainly wrong. Therefore, the check of first 160 bits of the key is enough to discard the majority of invalid candidate passwords [30].

8.3.2 HMAC Optimizations

[OPT–02] Block Reduction Since password p is an input parameter and it is not modified during the computation of PBKDF2, it is possible to precompute the first message block of a keyed hash function (light gray rectangles of Fig. 8.2) and reuse such a value in all the subsequent HMAC invocations. Thus, the number of blocks that have to be computed is reduced from "$4 * iteration\ count$" to "$2 + 2 * iteration\ count$." This simple optimization saves about 50% of PBKDF2's CPU-intensive operations [24, 30, 31].

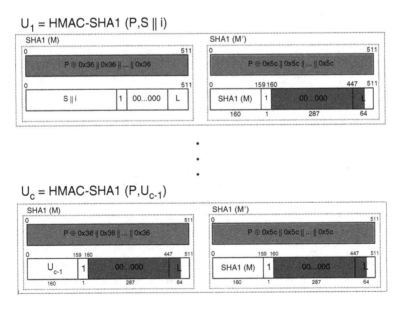

Fig. 8.2 PBKDF2-HMAC-SHA-1 optimizations

[OPT–03] Input Size A generic HMAC implementation has to address the problems of the size of password p and message $text$. If the password length is bigger than 512 bits, it has to be reduced. Therefore, a hash algorithm is applied, namely $p = SHA - 1(p)$, and then it is padded with enough zeros to reach a 512-bit length [32]. In addition, we have to address also the problem of the message size. If the message to be authenticated is bigger than 512 bits, it has to be split in several blocks and then each block managed separately. In PBKDF2, excluding the computation of U_1 (see Fig. 8.2), we have not a generic HMAC implementation but a specific one. Indeed, we know in advance the computation of the first message block (see [OPT–02] Merkle-Damgard block reduction), and we have to manage only the second one. Since the second message block always inputs a 160-bit message, namely $SHA-1(M)$ or U_i (see Fig. 8.2), we have not to split the message to be authenticated in blocks. Therefore, this optimization provides us the possibility to avoid length checks and the chunk splitting operations during the computation of $U_2,\ldots,U_c$, thus reducing the overhead necessary to compute an HMAC implementation [31].

8.3.3 SHA-1 Optimizations

[OPT–04] Word Expansion Phase Instead of using eighty 32-bit words for the word expansion phase (see Eq. (8.5)), SHA-1 can be implemented using a circular queue [26] of sixteen words (see Eq. (8.6)). This approach reduces the amount of memory required by the implementation, thus making this optimization a desirable feature in a GPU context.

A different approach has been introduced in [25], where the authors suggest the possibility to unfold the SHA-1 message scheduling function (see Eq. (8.9)). Although this approach increases the total number of XOR operations to be executed, it drastically reduces the amount of memory required to perform the SHA-1 message scheduling function, i.e., only five 32-bit words. Therefore, also this optimization may have an impact on GPU performances.

In addition, Visconti and Gorla [25] have also shown that Eq. (8.5) can be replaced with Eq. (8.7). This new approach does not reduce the amount of memory required to compute the word expansion phase but can be exploited to reduce the total number of XORs in a CPU context as suggested by [OPT–05].

[OPT–05] Zero-Based Optimization Due to a long run of several consecutive zeros, namely 287 bits, a number of 32-bit word W_t are set to zero. Since zero-based operations do not provide any contribution, they can be easily omitted. Therefore, exploiting Eqs. (8.7) and (8.8), we can avoid 66 out of 192 XOR operations [25].

The same approach can be adopted to reduce the number of constant XORed twice—i.e., 0x36 and 0x5C—when passwords p are short. We recall that XORing the same value twice does not provide any contribution and can be omitted [24].

[OPT–06] Three-Round Optimization During the computation of the message digest, SHA-1 operates on several 32-bit words such as constants K_t, registers A, B, C, D, E, functions f_t and W_t too. However, in the first three rounds a number of these words are known a priori and some operations can be omitted [31]. For example, in the first round we have to compute the following equation: $f_0 + E + ROTL(A, 5) + W_0 + K_0$ (see Fig. 8.1). The content of 32-bit word W_0 is unknown but those of f_0, E, the circular shift of A, and K_0 are not. Therefore, we can precompute $f_0 + E + ROTL(A, 5) + K_0 = 0x9FB498B3$ and reduce the first round to a single operation, namely $W_0 + 0x9FB498B3$, thus saving 3 operations out of 4. This approach can be also applied to second and third round, where the unknown values are A, W_1, and A, B, W_2, respectively.

8.4 Measuring Performances

To evaluate the contribution of the optimizations described in Sect. 8.3, we (a) implemented from scratch both CPU and GPU version of PBKDF2, (b) performed our testing activities, measuring PBKDF2 performances, and finally (c) compared our results with well-known implementations—e.g., OpenSSL version 1.1.0e [33], libgcrypt version 1.7.6 [34], hashcat 3.5.0 [35].

8.4.1 GPU Testing

In order to run the same code on several devices with different architectures, our implementation has been written using the OpenCL framework [36]. The implementation uses a classic host-device approach. The host (a CPU) generates a set of passwords and sends them to the device (a GPU). In order to exploit [OPT–03], our code executes PBKDF2 as a two-step process (see Fig. 8.3): firstly it computes U_1, storing the intermediate results (*hipad* and *hopad*) of the compression function— i.e., we are computing the light gray rectangles of Figs. 8.2 and 8.3—and secondly computes the remaining $U_2, \ldots, U_c$. Doing so, we can compute U_1 with a generic HMAC implementation and the remaining U_i with a specific one, thus avoiding length checks and the chunk splitting operations. In addition, after the computation of U_1, the CPU is able to transfer (asynchronously) a new set of candidate passwords to the GPU, reducing the overhead generated by read/write memory operations. All the tests were executed on a machine equipped with an AMD FX 8320 4 GHz processor, 8 GB RAM, Microsoft Windows 10 Home 64-bit operating system. In addition, we installed five consumer-grade GPUs with different architectures, memory structures, and price ranges: AMD R9 390, AMD HD6870, Nvidia GTX 960, Nvidia GTX 1060, and Nvidia GTX 670.

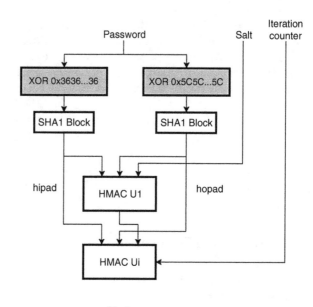

Fig. 8.3 PBKDF2 GPU implementation

Table 8.1 Number of kilohashes per second (KH/s) on different GPUs

GPU	Naive	[OPT–4] only	All SHA-1 opt.	All HMAC, SHA-1 opt.	Full version	hashcat
AMD R9 390	244.72	359.56	377.73	755.57	1553.34	1469.6
AMD HD6870	7.32	98.16	99.18	198.45	398.15	156.4
Nvidia GTX 670	75.28	84.14	90.06	191.20	393.83	410.7
Nvidia GTX 960	180.32	206.41	212.62	504.06	1048.44	992.2
Nvidia GTX 1060	324.26	351.64	381.34	1048.40	1678.30	1710.2

In order to show the contribution of the optimizations described in Sect. 8.3, we implement four different versions of our code:

1. based on [OPT–04];
2. based on all SHA-1 optimizations ([OPT–04], [OPT–05], [OPT–06]);
3. based on all HMAC and SHA-1 optimizations ([OPT–02], ..., [OPT–06]);
4. full version ([OPT–01], ..., [OPT–06]);

Then, we set the following PBKDF2 input parameters:

- iteration count $c = 1000$ (the minimum value suggested in [23]);
- derived key length $derkey = 256$ bits;
- a random salt s.

Finally, we run our implementations and collected data. Testing results are shown in Table 8.1 and Fig. 8.4. Then, we compare the performance of our code with a well-known password recovery utility [35].

8 Understanding Optimizations and Measuring Performances of PBKDF2

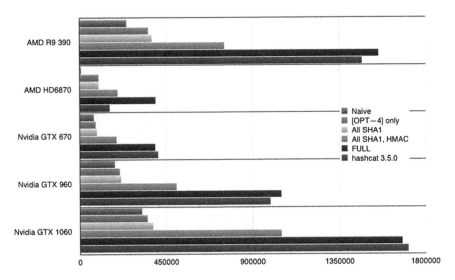

Fig. 8.4 Number of hashes per second on GPU

Table 8.2 Number of kilohashes per second (KH/s) on CPU

Library	Naive	[OPT–4] only	All SHA-1 opt.	All HMAC, SHA-1 opt.	Full version
Our version	0.797	0.974	1.005	1.761	1.791
OpenSSL ver.1.1.0e	–	–	–	–	0.851
Libgcrypt ver.1.7.6	–	–	–	–	0.683

8.4.2 CPU Testing

Since real-world applications may define an appropriate iteration count value at runtime by executing a CPU-test performance—e.g., LUKS [12, 37]—we also implemented a CPU-based version of PBKDF2. The main difference between a CPU and GPU implementation is that, in the first one, we have not to transfer a set of candidate passwords from host to device, hence we have not to split the algorithm in two phases as shown in Fig. 8.3. In addition, the CPU version implements Eqs. (8.7) and (8.8) as [OPT–4] instead of Eq. (8.6). In this case, the circular queue used to implement the word expansion phase of the GPU version does not provide any performance improvement. Indeed, a CPU-based approach has no memory constraints, while GPU-based has. Therefore, the circular queue is a desirable feature only in a GPU context.

Our testing activities were executed on an AMD FX-8320 8-Core 4 Ghz, setting iteration count c, derived key length $derkey$, and salt s as described in Sect. 8.4.1. Table 8.2 and Fig. 8.5 show the data collected by running our implementation, OpenSSL version 1.1.0e [33], and Libgcrypt ver.1.7.6 [34].

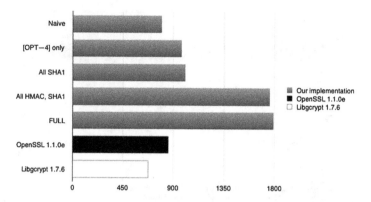

Fig. 8.5 Number of hashes per second on AMD FX 8230

8.5 Conclusions

User-chosen passwords are widely used to protect our sensitive information and to gain access to specific resources. They should be strong enough to prevent dictionary and brute-force attacks but usually are short and lack enough entropy and cannot be directly used as keys. A possible solution to these issues is to adopt a password-based key derivation functions. Although Argon2 [5], Catena [6], Lyra2 [7], yescrypt [8], Makwa [9], and scrypt [38] are expected to supersede PBKDF2 in the next years, currently PBKDF2 is still widely used to derive keys in many security applications. In this paper we described the state of the art of PBKDF2 with the aim to raise the reader's awareness of security issues facing new applications—for example, it is not difficult to find apps on Android market which implement PBKDF2 with poor security parameters. Thus, focusing on PBKDF2-HMAC-SHA-1, we described and tested a number of optimization techniques used by malicious users to speed up PBKDF2, HMAC, and SHA-1. In addition, we showed that these optimizations should be implemented in crypto libraries in order to speed up the performances, and accordingly, increasing the level of security of those applications which define the iteration count at runtime.

An interesting future work will be to analyze the performance of these optimizations on several graphics cards equipped with different chips from fastest to slowest.

References

1. C.E. Shannon, Prediction and entropy of printed English. Bell Syst. Tech. J. **30**(1), 50–64 (1951)
2. H. Krawczyk, Cryptographic extraction and key derivation: the HKDF scheme. Cryptology ePrint Archive. Report 2010/264 (2010)
3. K. Moriarty, B. Kaliski, A. Rusch, PKCS# 5: password-based cryptography specification version 2.1. RFC 8018 (2017)

4. Password hashing competition. https://password-hashing.net/. Accessed 10 Nov 2018
5. A. Biryukov, D. Dinu, D. Khovratovich, Argon2 (version 1.2). University of Luxembourg, Luxembourg. https://password-hashing.net/submissions/specs/Argon-v3.pdf. Accessed 10 Nov 2018
6. C. Forler, S. Lucks, J. Wenzel, Catena: a memory-consuming password-scrambling framework. Cryptology ePrint Archive. Report 2013/525 (2013)
7. M.A. Simplicio Jr., L.C. Almeida, E.R. Andrade, P.C. dos Santos, P.S. Barreto, Lyra2: password hashing scheme with improved security against time-memory trade-offs. Cryptology ePrint Archive. Report 2015/136 (2015)
8. A. Peslyak, yescrypt – password hashing scalable beyond bcrypt and scrypt. Openwall, Inc. (2014). http://www.openwall.com/presentations/PHDays2014-Yescrypt/. Accessed 10 Nov 2018
9. T. Pornin, The MAKWA password hashing function (2015). http://www.bolet.org/makwa/makwa-spec-20150422.pdf. Accessed 10 Nov 2018
10. Wi-Fi alliance: discover wi-fi: specifications. https://www.wi-fi.org/discover-wi-fi/specifications. Accessed 10 Nov 2018
11. iOS security guide (2017). https://www.apple.com/business/docs/iOS_Security_Guide.pdf. Accessed 10 Nov 2018
12. C. Fruhwirth, LUKS on-disk format specification version 1.2.2 (2016). https://gitlab.com/cryptsetup/cryptsetup/wikis/LUKS-standard/on-disk-format.pdf. Accessed 10 Nov 2018
13. A. Visconti, H. Tahayori, Detecting misbehaving nodes in MANET with an artificial immune system based on type-2 fuzzy sets, in *2009 International Conference for Internet Technology and Secured Transactions (ICITST)* (2009), pp. 1–2
14. M.T. Rahman, M.J.N. Mahi, Proposal for SZRP protocol with the establishment of the salted SHA-256 Bit HMAC PBKDF2 advance security system in a MANET, in *2014 International Conference on Electrical Engineering and Information Communication Technology* (2014), pp. 1–5
15. Enpass. https://www.enpass.io. Accessed 10 Nov 2018
16. F-secure key. https://www.f-secure.com/en/web/home_global/key. Accessed 10 Nov 2018
17. AgileBits: how PBKDF2 strengthens your master password. https://support.1password.com/pbkdf2/. Accessed 10 Nov 2018
18. LassPass: password iterations (PBKDF2). https://helpdesk.lastpass.com/account-settings/general/password-iterations-pbkdf2/. Accessed 10 Nov 2018
19. Keeper: keeper's best-in-class security. https://keepersecurity.com/security.html. Accessed 10 Nov 2018
20. A. Belenko, D. Sklyarov, "Secure Password Managers" and "Military-Grade Encryption" on Smartphones: Oh, Really? Blackhat Europe (2012)
21. L. Casati, A. Visconti, Exploiting a bad user practice to retrieve data leakage on android password managers, in *Proceedings of the 11th International Conference on Innovative Mobile and Internet Services in Ubiquitous Computing, IMIS 2017* (Springer, Berlin, 2017)
22. L. Casati, A. Visconti, The dangers of rooting: data leakage detection in android applications. Mob. Inf. Syst. **2018**, 6020461 (2018). https://doi.org/10.1155/2018/6020461
23. M.S. Turan, E.B. Barker, W.E. Burr, L. Chen, SP 800-132. Recommendation for password-based key derivation. Part 1: storage applications (2010). http://nvlpubs.nist.gov/nistpubs/Legacy/SP/nistspecialpublication800-132.pdf. Accessed 10 Nov 2018
24. A. Visconti, S. Bossi, H. Ragab, A. Caló, On the weaknesses of PBKDF2, in *Proceedings of the 14th International Conference on Cryptology and Network Security, CANS 2015*. Lecture Notes in Computer Science, vol. 9476 (Springer, Berlin, 2015)
25. A. Visconti, F. Gorla, Exploiting an HMAC-SHA-1 optimization to speed up PBKDF2. IEEE Trans. Dependable Secure Comput. (2018). https://doi.org/10.1109/TDSC.2018.2878697
26. NIST: FIPS PUB 180-4. Secure Hash Standard (SHS) (2012). http://nvlpubs.nist.gov/nistpubs/FIPS/NIST.FIPS.180-4.pdf. Accessed 10 Nov 2018
27. M. Bellare, R. Canetti, H. Krawczyk, Keying hash functions for message authentication, in *Proceedings of Advances in Cryptology—CRYPTO96* (Springer, Berlin, 1996), pp. 1–15

28. M. Bellare, R. Canetti, H. Krawczyk, Message authentication using hash functions—the HMAC construction. RSA Lab. CryptoBytes **2**(1), 12–15 (1996)
29. H. Krawczyk, M. Bellare, R. Canetti, HMAC: keyed-hashing for message authentication. RFC 2104
30. A. Ruddick, J. Yan, Acceleration attacks on PBKDF2: or, what is inside the black-box of oclHashcat? in *Proceedings of the 10th USENIX Workshop on Offensive Technologies* (2016)
31. J. Steube, Optimising computation of hash-algorithms as an attacker. https://hashcat.net/events/p13/js-ocohaaaa.pdf. Accessed 10 Nov 2018
32. NIST: FIPS PUB 198-1. The keyed-hash message authentication code (HMAC) (2008). http://csrc.nist.gov/publications/fips/fips198-1/FIPS-198-1_final.pdf. Accessed 10 Nov 2018
33. Openssl, version: 1.1.0e. https://www.openssl.org/. Accessed 10 Nov 2018
34. Libgcrypt, version 1.7.6. https://www.gnupg.org/software/libgcrypt/index.html. Accessed 10 Nov 2018
35. hashcat, version 3.30. https://hashcat.net/hashcat/. Accessed 10 Nov 2018
36. OpenCL. https://www.khronos.org/opencl/. Accessed 10 Nov 2018
37. S. Bossi, A. Visconti, What users should know about full disk encryption based on LUKS, in *Proceedings of the 14th International Conference on Cryptology and Network Security, CANS 2015*. Lecture Notes in Computer Science, vol. 9476 (Springer, Berlin, 2015)
38. C. Percival, Stronger key derivation via sequential memory-hard functions (2009). https://www.tarsnap.com/scrypt/scrypt.pdf. Accessed 10 Nov 2018

Chapter 9
PSARV: Particle Swarm Angular Routing in Vehicular Ad Hoc Networks

Mrigali Gupta, Nakul Sabharwal, Priyanka Singla, Jagdeep Singh, and Joel J. P. C. Rodrigues

9.1 Introduction

Vehicular ad hoc networks (VANET) [1] are a subclass of mobile ad hoc networks [2], where nodes are vehicles moving freely and providing network functionality. The architecture of a VANET consists of in-vehicle domain that uses one or more application units (AUs), road-side unit (RSU), on-board unit (OBU), ad hoc domain consisting of OBUs, and an alongside road-side unit and an infrastructure domain connected to the Internet. In this architecture of message passing between vehicle to vehicle and vehicle-to-roadside units using wireless links, the challenge from a route discovery standpoint is to establish a reliable routing path to carry the message from source and destination nodes, accounting for rapid changing topology due to nodes' mobility, driver's behavior, lack of fixed access point, limited node's memory, performance degradation on long distance between source and destination, increased network congestion, to name a few [3]. Existing literature on routing in VANETs has shown that traditional routing protocols for MANETs based on network topology do not necessary apply to VANETs; consequently, many position-based routing protocols that exploit the navigation information for

---

M. Gupta · N. Sabharwal · P. Singla · J. Singh (✉)
Division of Information Technology, Netaji Subhas Institute of Technology, University of Delhi, New Delhi, India

J. J. P. C. Rodrigues
National Institute of Telecommunications (Inatel), Santa Rita do Sapucaí, MG, Brazil

Instituto de Telecomunicações, Lisboa, Portugal

University of Fortaleza (UNIFOR), Fortaleza, CE, Brazil
e-mail: Joeljr@ieee.org

© Springer Nature Switzerland AG 2019
I. Woungang, S. K. Dhurandher (eds.), *2nd International Conference on Wireless Intelligent and Distributed Environment for Communication*, Lecture Notes on Data Engineering and Communications Technologies 27, https://doi.org/10.1007/978-3-030-11437-4_9

VANETs have been proposed, each of which utilizes its own approach to deal with one or many of the aforementioned constraints [4], while achieving timely, safe, and reliable dissemination of messages between nodes. Swarm intelligent-based routing protocols [5] are a subclass of such protocols. Unlike recent swarm intelligent-based routing protocols for VANETs [6], this paper introduces a novel PSO-based routing protocol (called particle swarm angular routing in VANETs (PSARV)), which uses the PSO technique as backbone of angular routing to determine a suitable routing path to carry a message from source to destination. To do so, each route request packet is treated as a swarm, which in turn calculates the next hop of a node using the PSO technique. The proposed PSARV scheme is evaluated by simulations and its performance is compared against that of the DSR protocol using packet delivery ratio, lost packet percentage, and throughput, as performance metrics. The rest of the paper is organized as follows: In Sect. 9.2, a brief overview of DSR protocol is presented. In Sect. 9.3, the design of the PSARV protocol is described in-depth. In Sect. 9.4, simulation results are discussed. Finally, Sect. 9.5 concludes the paper.

9.2 Dynamic Source Routing

The DSR protocol [7] is a reactive protocol that uses a multi-hop approach for transmitting the data from one node to another. In this procedure, the *route discovery* consists in finding the duplex routes from source to destination with a combined effort of all intermediate nodes. Assume that a node, say $n1$, wishes to send a packet to another node, say $n2$. In this setting, *RREQ* packets are flooded in the network by node $n1$ if no route to $n2$ already exists in node $n1'$ cache. Upon receiving the *RREQ*, any intermediate node, say $n3$, will unicast a *ROUTE REPLY (RREP)* to $n1$ if it knows a valid route to $n2$ along with the route information; otherwise, the *RREQ* is transmitted forward. In other words, a *RREP* substantiates the existence of a route, and this newly discovered route is added to the cache of all nodes present in it. On the other hand, the *route maintenance* procedure is invoked to detect any changes that might have occurred in the network topology. In the above procedures, if a packet needs to be sent from $n1$ to $n2$ and the aforementioned route cached turns out to be invalid, any other available valid route will be utilized or the route discovery procedure will be invoked again. As a table-driven protocol, DSR yields low bandwidth and power consumption, as well as reduced network traffic since no table update is required and the cache information is used to efficiently reduce the control overheads. Some of the deficiencies of the DSR protocol lie in the fact that its route discovery procedure is initiated for every message to be transmitted separately, leading to a substantial connection setup delay. Also, increased error rate may occurred due to the lack of a step for checking the existence of a route before sending the data packets, which may lead to some inconsistency in the DSR packet forwarding mechanism. As will be shown in the simulation results, when utilized as a VANET protocol, the DSR performance degrades gradually in highly mobile scenarios.

9.3 Design of the PSARV Protocol

The proposed PSARV protocol is designed to address the above shortcomings of the DSR protocol in the VANET environment by avoiding broadcasting by intermediate nodes and by reducing the number of control packets. In PSARV, each route request packet is considered as a single swarm that has velocity, position coordinates, personal best experience value, personal best experience position, global best experience position or destination, time-to-live (TTL), source address, and destination address. The next hop of a swarm is calculated using PSO equations involving the current velocity, current position coordinates, personal best experience, personal best experience position, and global best experience coordinates.

9.3.1 Swarm Movement Algorithm

Each RREQ-swarm packet contains the source node address, destination node address—which is the global optimum, personal best experience value—initially set to cost(source.position) using Eq. (9.3), personal best experience position—initially set to source.position, velocity—initially set to 0, TTL, last node address, inertia factor value—set to a suitable value, acceleration factor related to global best—set to a suitable value, acceleration factor related to personal best—set to a suitable value, and r_1 and r_2 —which are random numbers in between 0 and 1.

The source node broadcasts RREQ-swarms to its neighbors (as illustrated in Fig. 9.1). These RREQ-swarms are responsible to find a suitable route from source to destination in order to transmit the data. Once any RREQ-swarm reaches the destination, it sends back a route reply message to the source. This route reply

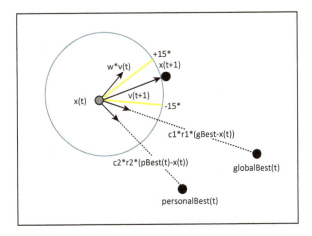

Fig. 9.1 RREQ-swarm forwarding

message uses an entry in the routing table which has already been updated by the RREQ-swarm. In doing so, the PSO equations that are involved are as follows:

$$v(t+1) = w * v(t) + c_1 * r_1 * (pB(t) - x(t)) + c_2 * r_2 * (gB(t) - x(t)) \quad (9.1)$$

$$x(t+1) = x(t) + v(t+1) \quad (9.2)$$

$$cost(x) = |gB(t) - x(t)|^2 \quad (9.3)$$

where $x(t)$ is the position of the particle at time t, w is the inertia factor, c_1 and c_2 are the acceleration factors related to global best and local-best, respectively, $v(t)$ is the particle velocity at time t, r_1 and r_2 are the aforementioned random numbers, $pB(t)$ (or $pBest(t)$) are the personal best experience coordinates, pBestValue is the personal best experience value, $gB(t)$ (or $gBest(t)$) is the global best (or destination) coordinates, and $cost(x)$ is the cost function (in terms of distance between current position and destination squared). Assuming that (1) every node can extract the destination coordinates using GPS; (2) every node can access the location lookup table which stores the node's position, velocity, and ID; (3) all the nodes are moving at a moderate speed; and (4) all nodes willingly participate in establishing the source to destination path, the steps of the swarm movement algorithm are described in Algorithm 1.

9.3.2 PSARV Protocol

The main steps of the PSARV protocol design are described as follows (in the order shown): (1) Initiate the route discovery; (2) Move the received RREQ-swarm forward; (3) Initiate a route retry; (4) Initiate a route reply; and (5) Start the data transmission. The details are provided in the sequel.

9.3.2.1 Initiate the Route Discovery

This procedure (depicted in Fig. 9.2) works as follows:

1. If the source node has a fresh route in its routing table, use that entry.
2. Else generate the RREQ-swarms and set their content parameters to initial values. Then, broadcast the RREQ-swarms only to its nearby neighbors (as shown in Fig. 9.2). These RREQ-swarms will go on discovery to find a suitable route from source to destination to be used for data transmission.

9 PSARV: Particle Swarm Angular Routing in Vehicular Ad Hoc Networks

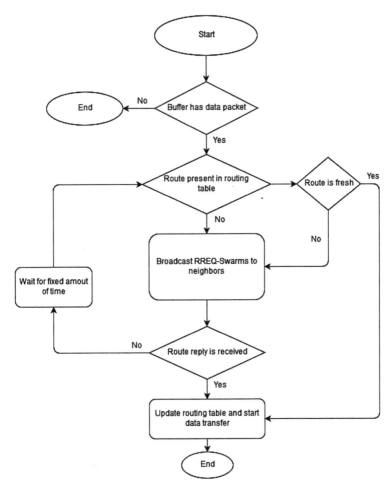

Fig. 9.2 Sender route request initiation

9.3.2.2 Moving the Received RREQ-Swarm Forward

This procedure (depicted in Fig. 9.3) works as follows:

1. If RREQ-swarm is duplicate, discard it.
2. If it is received by the initiator of swarm, discard it as well.
3. If RREQ-swarm is not a duplicate, insert its broadcast ID into the lookup seen table.
4. Update the neighbor table, seen table, and routing table for the current node.

Fig. 9.3 RREQ-swarm forwarding by intermediate nodes

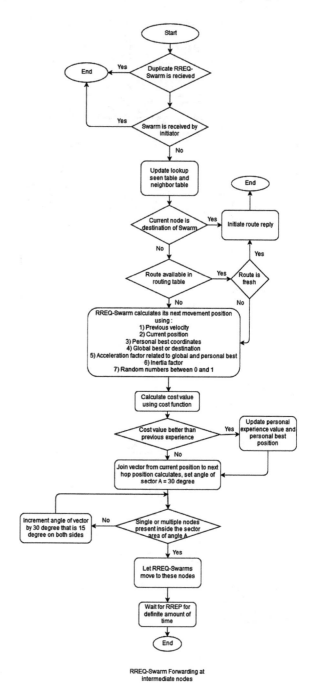

9 PSARV: Particle Swarm Angular Routing in Vehicular Ad Hoc Networks

---

Algorithm 1 Swarm movement algorithm

1: **procedure** MOVESWARMFORWARD($RREQ - swarm, node$)   ▷ RREQ-swarm movement from node

2:    **Update** routing table using RREQ-swarm.last-address.

3:    $RREQ - swarm.last - address \leftarrow node.name$

4:    **Update** neighbor table using RREQ-swarm.last-address.

5:    **Update** seen table.

6:    $gB(t) \leftarrow RREQ - swarm.destination$

7:    $pB(t) \leftarrow RREQ - swarm.personal Best Position$

8:    $v(t) \leftarrow RREQ - swarm.velocity$

9:    $x(t) \leftarrow node.position$

10:    $pBestValue \leftarrow RREQ - swarm.pBestValue$

11:    $\mathbf{v(t+1)} \leftarrow w * v(t) + c_1 * r_1 * (pB(t) - x(t)) + c_2 * r_2 * (gB(t) - x(t))$

12:    $RREQ - swarm.velocity \leftarrow v(t + 1)$

13:    $\mathbf{x(t+1)} \leftarrow x(t) + v(t + 1)$

14:    $\mathbf{cost(t)} \leftarrow |gB(t) - x(t)|^2$

15:    **if** cost(x) > pBestValue **then**

       $RREQ - swarm.personal Best Position \leftarrow x(t)\ RREQ - swarm.pBestValue \leftarrow cost(x)$

16:    $\theta \leftarrow direction\ of\ \overrightarrow{x(t+1)}\ with\ X - axis$

17:    $ang \leftarrow 15°$

18:    **while** $\theta + ang < 180°$ **do**

19:       **if** node is present in $\theta - ang$ or $\theta + ang$ **then** break.

20:       **else** $ang \leftarrow ang + 15°$

21:    **end while**

22:    Forward RREQ-swarm to all nodes in this sector.

23:    Done.

24: **end procedure**

---

5. If current node is the destination of the RREQ-swarm, update the routing table and initiate a RREP message.

6. If current node is an intermediate node and a fresh route is available, then initiate a route reply (RREP) message.

7. If current node is an intermediate node and no route is available, move the RREQ-swarm forward (as depicted in Fig. 9.3) using the aforementioned swarm movement algorithm (Algorithm 1).

8. If current node is an intermediate, then the available route is not fresh. In this case, move RREQ-swarm forward (as depicted in Fig. 9.3) using the aforementioned swarm movement algorithm (Algorithm 1).

9. If the number of broadcasts exceeds the maximum number allowed, generate a RERR message and start the *Initiate the Route Discovery* process again.

Route Retry This step occurs whenever there is timeout at the intermediate or source node. In such case, the route retry phase consists of re-generating the RREQ-swarms by means of the *Initiate the Route Discovery* phase.

Route Reply This step applies in the following scenarios: (1) if a node address is that of the destination of the RREQ-swarm, then initiate a route reply; (2) if the current node is an intermediate node (i.e., between source and destination), then relay the route reply message up to the source node by using the entry in the routing table that was updated by the RREQ-swarm during the *Initiate the Route Discovery* phase.

Start the Data Transmission This step consists of transferring the data packets via the discovered reliable route until all they are finished.

9.4 Performance Evaluation

The proposed PSARV protocol is evaluated by simulations using GloMoSim [8] and compared against the DSR protocol in VANET environment.

9.4.1 Simulation Parameters

The simulation parameters are captured in Table 9.1.

The considered performance metrics are:

1. Throughput: Number of bits successfully delivered from source to destination per unit time (bits/second).
2. Packet delivery ratio (PDR): Number of packets received by destination divided by the number of packets generated by the source node.
3. Loss packet percentage: Percentage of packets lost divided by the packets generated by the sender.
4. End-to-end delay: Time taken (in milliseconds) for a packet to get delivered from source to destination.

Table 9.1 Simulation settings

Name	Value
Terrain dimension	$(2000 \times 2000) \, m^2$
Maximum speed	0–50 m/s
Total simulation time	15 min
Mobility	Random waypoint
Number of nodes	Up to 65
Packet size	512
Node placement	Uniform
Range of transmission	376.782 m
Medium access control protocol	IEEE 802.11
Network protocol	IP
Propagation limit	−111.0 decibel
Radio type	Radio Accnoise
Radio frequency	2.4 GHz

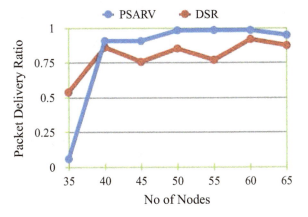

Fig. 9.4 Packet delivery ratio v/s no. of nodes

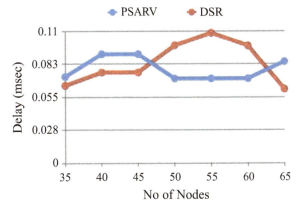

Fig. 9.5 Average delay v/s no. of nodes

9.4.2 Simulation Results

The first set of results (Figs. 9.4, 9.5, 9.6, and 9.7) has been obtained using static nodes, whereas the second set of results (Figs. 9.8, 9.9, and 9.10) has been obtained using mobile nodes.

First, the number of nodes is varied and the impact of this variation on the PDR, end-to-end delay, throughput, and loss packet percentage are evaluated. The results are captured in Figs. 9.4, 9.5, 9.6, and 9.7, respectively.

In Fig. 9.4, it can be observed that initially, when the number of nodes is less than 40, the PDR ratio obtained using DSR is higher than that obtained using PSARV, but as soon as the number of nodes surpasses that threshold of 40 (which is often the case in a VANET environment), PSARV becomes more efficient than DSR. In some instances, the PDR ratio obtained using the PSARV scheme is almost equal to 1. This is attributed to the fact that in the PSARV scheme, the RREQ-swarms are designed to search for a suitable reliable routing path to destination over a large geographical area, which is a more natural way, whereas in the DSR scheme, the

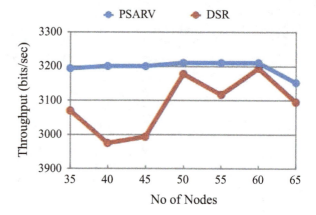

Fig. 9.6 Throughput v/s no. of nodes

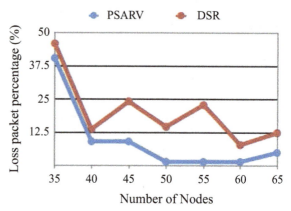

Fig. 9.7 Loss packet percentage v/s no. of nodes

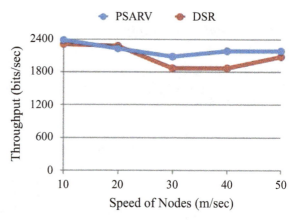

Fig. 9.8 Throughput vs. speed of nodes

route may be found by only broadcasting the route request to neighbors. In PSARV, the swarm movement at each step takes into account the destination node's position, whereas in DSR, there is no such consideration.

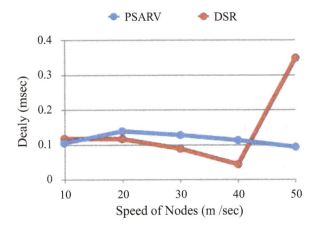

Fig. 9.9 Average delay vs. speed of nodes

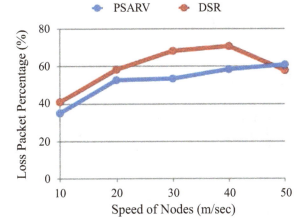

Fig. 9.10 Loss packet percentage vs. speed of nodes

In Fig. 9.5, it can be observed that under a certain threshold in the number of nodes (here near 50), the end-to-end delay incurred by packets when using DSR is less than that obtained using PSARV; but as soon as that threshold is surpassed, the end-to-end delay generated by PSARV often decreases significantly compared to that generated by DSR. Again, this may be attributed to the implemented RREQ-swarm search strategy.

In Fig. 9.6, it can be observed that independently of the increase in the number of nodes, the throughput generated by the PSARV scheme is always significantly higher than that generated by the DSR scheme. Consequently, the loss packet percentage is much higher when using PSARV compared to DSR (Fig. 9.7). For a higher number of nodes, the loss packet percentage drop is more significant when using PSARV than DSR.

Second, the node's speed is varied and the impact of this variation on the throughput, end-to-end delay, and loss packet percentage are evaluated, when the number of nodes is set to 90. The results are captured in Figs. 9.8, 9.9, and 9.10, respectively.

In Fig. 9.8, it can be observed that initially, when the node's speed is low, the throughput obtained using PSARV and DSR is almost similar, but as soon as the node's speed surpasses the threshold of 20, the throughput obtained using PSARV becomes much higher than that obtained using DSR. This is attributed to the route discovery and establishment process implemented in PSARV, which takes into account the position of the destination node and which does not rely on table entries.

In Fig. 9.9, it can be observed that the end-to-end delay generated by the PSARV scheme remains consistent when the node's speed increases even though it is a bit higher for low node's speed (up to a threshold of 40); but then the end-to-end delay for DSR shoots up for higher node's speed (>40) compared to that obtained for PSARV, which in fact remains consistent.

In Fig. 9.10, it can be observed that the PSARV scheme performs more efficiently than the DSR scheme. Initially, PSARV does not exhibit a noticeable packet drop compared to DSR, but as the node's speed increases, the drop becomes significantly higher. This might be attributed to the route discovery and establishment process used in PSARV.

9.5 Conclusion

In this paper, we have proposed an efficient particle swarm angular routing protocol (called PSARV) for VANETs which uses a PSO technique as backbone of angular routing to find a suitable path to route a message from source to destination. In this protocol, the RREQ-swarms are designed to accomplish such search. Simulations results have shown that: (1) in dense traffic conditions, PSARV outperforms DSR in terms of throughput, lost packet percentage, and packet delivery ratio; (2) when the number of nodes is fixed to 90 and the node's speed is increased, PSARV generally performs more efficiently than DSR in terms of throughput and loss packet percentage. In terms of end-to-end delay, the PSARV scheme remains consistent in dense traffic conditions when the node's speed increases. As future work, the proposed swarm movement algorithm can be utilized in Platoon leader swarm-based multicasting for the safety of passengers. In this case, when the Platoon leader detects any danger in its lane, a multicast signal is triggered and the RREQ-swarms are sent along with the function having the minima at the desired lane. This way, all vehicles in that lane are warned of the danger ahead so that drivers are already alerted and awake. As another future work, we also intend to evaluate the performance of PSARV in real traffic network scenarios.

Acknowledgements This work is partially supported by a grant from the National Funding from the FCT—Fundação para a Ciência e a Tecnologia through the UID/EEA/500008/2013 Project, by Finep, with resources from Funttel, Grant No. 01.14.0231.00, under the Radiocommunication Reference Center (Centro de Referência em Radiocomunicações, CRR) Project of the National Institute of Telecommunications (Instituto Nacional de Telecomunicações, Inatel), Brazil, by Brazilian National Council for Research and Development (CNPq) via Grant No. 309335/2017-5, all held the fifth author.

References

1. S. Olariu, M.C. Weigle, *Vehicular Networks: From Theory to Practice*. Chapman & Hall/CRC Computer and Information Science Series (Chapman & Hall/CRC, Boca Raton, 2017), 472 pp. ISBN 9781138116597
2. S.K. Dhurandher, G.V. Singh, Weight based adaptive clustering in wireless ad hoc networks, in *Proceedings of IEEE International Conference on Personal Wireless Communication (ICPWC 2005)*, New Delhi, January 2005 (IEEE, Piscataway, 2005), pp. 95–100
3. S. Harrabia, W. Chainbi, K. Ghedira, A multi-agent approach for routing on vehicular ad-hoc networks, in *Proceedings of the 4th International Conference on Ambient Systems, Networks and Technologies (ANT 2013)*. Procedia Computer Science, vol. 19 (2013), pp. 578–585
4. R. Brendha, V.S.J. Prakash, A survey on routing protocols for vehicular ad hoc networks, in *Proceedings of the 4th IEEE International Conference on Advanced Computing and Communication Systems (ICACCS)*, Coimbatore, January 6–7, 2017
5. S. Bitam, A. Mellouk, S. Zeadally, Bio-inspired routing algorithms survey for vehicular ad hoc networks. IEEE Commun. Surv. Tutor. **17**, 843–867 (2015)
6. R.C. Poonia, A performance evaluation of routing protocols for vehicular ad hoc networks with swarm intelligence. Int. J. Syst. Assur. Eng. Manag. (2017). https://doi.org/10.1007/s13198-017-0661-1 (Last visited April 9, 2018)
7. D.B. Johnson, D.A. Maltz, Dynamic source routing in ad wireless networks, in *Mobile Computing*, Ch. 5, ed. by T. Imielinski, H. Korth (Kluwer Academic Publishing, Norwell, 1996), pp. 153–181
8. GlomoSim: Global Mobile Information Systems Simulation Library. https://pcl.cs.ucla.edu/project/glomosim (Last visited April 9, 2018)

Chapter 10
A Reliable Firefly-Based Routing Protocol for Efficient Communication in Vehicular Ad Hoc Networks

Nakul Sabharwal, Priyanka Singla, Mrigali Gupta, Jagdeep Singh, and Joel J. P. C. Rodrigues

10.1 Introduction

Vehicular ad hoc networks (VANETs) [1] are a subclass of mobile ad hoc networks (MANETs) [2] whose communication infrastructure is made of road side, on-board, and application units, which are responsible for message passing and Internet connection of vehicles. These types of networks are mainly characterized by instant connection establishment between nodes; and with respect to messages transfer, it is advocated that on-demand routing protocols be utilized as opposed to static primitive protocols [1]. Following a typical architecture of VANETs [1] (depicted in Fig. 10.1), in order to design a reliable routing protocol for VANETs, one should account for the rapid changing VANET's topology due to mobility of nodes, the absence of fixed access point, the node's memory limitation, to name a few [3]. Considering these requirements, this paper focuses on an approach where a swarm-intelligent algorithm (called firefly-swarm algorithm) is proposed to achieve the route request message passing. Swarm-intelligent-based routing schemes [4] are a subclass of meta-heuristic nature inspired algorithms [5] which have been used for global optimization of complex functions. These algorithms work with backbone

---

N. Sabharwal · P. Singla · M. Gupta · J. Singh (✉)
Division of Information Technology, Netaji Subhas Institute of Technology, University of Delhi, New Delhi, India

J. J. P. C. Rodrigues
National Institute of Telecommunications (Inatel), Santa Rita do Sapucaí, MG, Brazil

Instituto de Telecomunicações, Lisboa, Portugal

University of Fortaleza (UNIFOR), Fortaleza, CE, Brazil
e-mail: Joeljr@ieee.org

© Springer Nature Switzerland AG 2019
I. Woungang, S. K. Dhurandher (eds.), *2nd International Conference on Wireless Intelligent and Distributed Environment for Communication*, Lecture Notes on Data Engineering and Communications Technologies 27,
https://doi.org/10.1007/978-3-030-11437-4_10

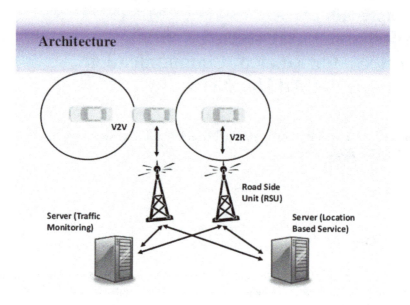

Fig. 10.1 VANET architecture

based on biological cooperative insects, physical and chemical systems. Firefly is one of the meta-heuristic algorithms which uses twinkling behavior of fireflies for finding global optimum [4]. In this paper, a reactive routing protocol for VANETs (called efficient firefly routing protocol (EFR for short)) is proposed to discover a suitable route from source to destination for message passing. The proposed EFR scheme consists of a firefly optimization algorithm with angular routing, which uses social interaction among fireflies to determine the best route for message passing from source to destination. The performance of the proposed EFR protocol is evaluated by simulations using GloMoSim [6], showing that it outperforms the DSR routing protocol in terms of throughput, packet delivery ratio, lost packet percentage, and end-to-end delay, both under static and high mobility scenarios.

The rest of the paper is organized as follows: In Sect. 10.2, some background and related work are presented. In Sect. 10.3, the design of the EFR protocol is described in-depth. In Sect. 10.4, simulation results are presented. Finally, Sect. 10.5 concludes the paper.

10.2 Background and Related Work

This section provides an overview of the firefly and DSR algorithms. A discussion on swarm-intelligent-based routing algorithms for VANETs is also provided.

10.2.1 Firefly Algorithm

The main principles of the firefly algorithm (FA) [7] are as follows: (1) Fireflies are attracted by light; they move toward each other; and the attraction is not based on gender; (2) the dimmer fireflies go toward brighter ones; and (3) some randomness in motion prevails. Following these principles, the so-called firefly algorithm has been proposed in [8], taking the following parameters into account:

- Attractiveness of the firefly: This is directly proportional to the brightness of the firefly and is indirectly proportional to the distance between them [8], i.e.,

$$\text{Attractiveness} \propto \text{Brightness} \propto 1/\text{Distance}. \qquad (10.1)$$

- Firefly's light intensity I. This is obtained [8] as

$$I = I_0 e^{-\gamma r^2}, \qquad (10.2)$$

where r is the distance, γ is the light absorption coefficient, and I_0 is the light intensity at $r = 0$.
- Firefly's attractiveness β. This is obtained [8] as

$$\beta = \beta_0 e^{-\gamma r^2}, \qquad (10.3)$$

where β_0 is the attraction coefficient at $r = 0$.
- Distance between two fireflies at x_i and x_j. This is obtained [8] as

$$r_{ij} = ||x_i - x_j|| = \sqrt[2]{\sum_{k=1}^{d}(x_i^k - x_j^k)^2}. \qquad (10.4)$$

- Motion of a dimmer firefly i toward a dazzling firefly j. This is obtained [8] as

$$x_i^{t+1} = x_i^t + \beta_0 e^{-\gamma r_{ij}^2}(x_j^t - x_i^t) + \alpha_t \epsilon_i^t, \qquad (10.5)$$

where α is the maximum radius of the random step, β is the step size toward the optimal solution, ϵ_i is a uniformly distributed random number, x_i^t is the current position of ith firefly, $j > 1$, x_j^t is the current position of jth firefly, x_i^{t+1} is the next hop position of the firefly, α is the mutation coefficient, δ is the uniform mutation range (set at 100 in our simulations), r is the distance from destination, and r_{ij} is the distance from the ith node to the jth node.

10.2.2 Dynamic Source Routing

The dynamic source routing (DSR) [9] is a reactive protocol that uses a multi-hop technique for the transmission of data. In its *route discovery* phase, alternate routes from source to destination are discovered with the help of all intermediate nodes. For instance, if a node, say $n1$, wishes to transmit a packet to node $n2$, it will flood the network with route request (*RREQ*) packets. Upon receiving the *RREQ*, any intermediate node, say n_i, will unicast a route reply *RREP* to node $n1$ if it knows a valid route to $n2$ and this route will be added to the cache of all nodes present in it; otherwise, the *RREQ* will be transmitted forward and the route discovery procedure will be restarted again. In the *route maintenance* phase, any change that may have occurred in the network topology will be detected. The advantages of DSR over table-driven protocols stem from its low bandwidth consumption as well as low power consumption. On the other hand, the route discovery process in DSR is initiated anytime that a new message is to be transmitted, which may lead to considerable routing overhead caused by the incurred connection setup delay, inconsistency in the packet forwarding mechanism, and gradual decay of network performance in highly mobile scenarios (such as in VANET's application scenarios).

10.2.3 Related Work

A comprehensive survey on routing protocols for VANETs is given in [10]. Among these is the subclass of swarm-intelligent-based algorithms [4], to which the firefly algorithm belongs. Owing to its nature inspired feature, this type of algorithms uses its own heuristic to search for a suitable route to transfer the data packets from source to destination. Representative such algorithms have been discussed in [4, 11, 12]. In [4], a taxonomy of bio-inspired routing protocols for VANETs is proposed. In [11], Chhabra and Kumar proposed a routing technique based on the idea of firefly algorithm to achieve efficient packets transfer from source vehicle to destination vehicle in VANETs. The proposed technique is shown to yield less overhead and lower transmission time. In [12], Oranj et al. proposed a routing protocol for VANETs based on ant colony optimization and the dynamic MANET on-demand protocol. In this scheme, the route maintenance process is meant for eliminating the invalid paths from the routing tables according to a duration time limit imposed on the construction of each path.

10.3 Proposed Firefly Routing Protocol Protocol

In this section, the design of the proposed firefly routing protocol (EFR) for VANETs is described as a two-step process: Step 1—which consists of the

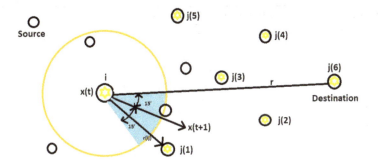

Fig. 10.2 Firefly-swarm next hop selection

route discovery process (achieved using the Firefly-Swarm Next Hop Selection Algorithm) and Step 2—which consists of the EPR technique itself.

10.3.1 EFR Route Discovery

The proposed firefly-swarm next hop selection mechanism is illustrated in Fig. 10.2, where each route request packet is considered as a firefly-swarm that has position coordinates, destination coordinates, time-to-live (TTL), and source and destination addresses. The next hop of the firefly-swarm is computed by using the firefly optimization equations [8] involving the current firefly coordinates, destination coordinates, brightness and attractiveness of other fireflies, brightness of destination, and angle between the neighbor nodes' coordinates.

For design purpose, the following assumptions are considered: (1) the destination's position coordinates of the firefly are available using GPS; (2) each node has permission and can access the RSU-firefly-database that contains each firefly's current coordinates; and (3) each node is readily engaged in setting up the source to destination route. Based on this, each firefly-swarm works according to the following procedure (referred to as *PROCEDURAL OPERATIONS*):

1. Each firefly-swarm is supposed to carry the following information: address of source node, address of destination node (which represents the brightest firefly and global optimum), TTL, last node address, light absorption coefficient γ, attraction coefficient base value β, mutation coefficient α, and destination coordinates.
2. The source node generates an initial population of firefly-swarms and sends it to every neighbor. These firefly-swarms will be responsible for discovering the suitable route as per Algorithm 1 described below, taking into account the parameters defined in the previously described Eqs. (10.2)–(10.5).

Algorithm 1 Firefly-swarm next hop selection algorithm

1: **procedure** FIREFLYSWARMNEXTHOP($firefly - swarm, currentNode$)   ▷ Selection of next node by firefly-swarm
2:     **Update** Put firefly-swarm.last-address in routing table.
3:     $firefly - swarm.last - address \leftarrow currentNode.number$
4:     **Update** Put firefly-swarm.last-address in neighbor table.
5:     **Update** Put firefly-swarm in seen table.
6:     **Update** Update central database with firefly-swarm and its position using road side units.
7:     $dest \leftarrow firefly - swarm.destination$
8:     $x_i(t) \leftarrow currentNode.coordinates$
9:     Access every other firefly location using road side unit's central database
10:     $fireflies - list \leftarrow RSU - firefly - database.getList(destination)$ ▷ getList function returns list of name and coordinates of fireflies.
11:     fireflies-list.append(dest)
12:     **for** i:1 to fireflies-list.length() **do**
13:         $r_j(t) \leftarrow$ fireflies-list[i].coordinates
14:         $r_{ij} \leftarrow ||r_j(t) - r_i(t)||$
15:         $I_i \leftarrow I_0 e^{-\gamma(||r_i(t)-dest||)^2}$
16:         $I_j \leftarrow I_0 e^{-\gamma(||r_j(t)-dest||)^2}$
17:         $\beta \leftarrow \beta_0 e^{-\gamma r_{ij}^2}$
18:         **if** $I_j > I_i$ **then**
19:             $m \leftarrow 2$
20:             $\beta \leftarrow \beta_0 * e^{-\gamma * r_{ij}^m}$
21:             $\epsilon \leftarrow \delta * unifrnd(-1, +1)$
22:             $x_i^{t+1} \leftarrow x_i^t + \beta * rand() * (x_j^t - x_i^t) + \alpha * \epsilon^t$
23:     **end for**
24:     $\theta \leftarrow direction\ of\ \overrightarrow{x_i(t+1)}\ with\ X - axis$
25:     $angVar \leftarrow 15°$
26:     **while** $\theta + angVar <= 180°$ **do**
27:         **if** $node \in [\theta - angVar, \theta + angVar]$ **then** break.
28:         **else** $angVar \leftarrow angVar + 15°$
29:     **end while**
30:     Send firefly-swarm to nodes in this sector.
31:     Function ends.
32: **end procedure**

10.3.2 EFR Technique

The steps of this technique are described as follows:

1. Initiate the route discovery mechanism:

   (a) If there is a fresh tuple in the routing table, then use it.
   (b) Else create firefly-swarms and initialize their contents as described earlier in *PROCEDURAL OPERATIONS*. Firefly-swarms now start the exploration of the route from source to destination.

2. Perform the following tasks when a firefly-swarm is received:

   (a) If the firefly-swarm is redundant, do nothing.
   (b) If the firefly-swarm is received by the sender itself, do nothing.

10 A Reliable Firefly-Based Routing Protocol for Efficient Communication. . .   135

(c) If the firefly-swarm is a newly arrived one, then put its broadcast-ID into the lookup seen table.
(d) Update the routing table and neighbor table using lastAddress content of the firefly-swarm.
(e) If the firefly-swarm has reached the destination, update the routing table and start a route reply (RREP).
(f) If current node is a midway node, and the latest path is present in the routing table, then start a route reply (RREP).
(g) If current node is a midway node, and no route is present in the routing table, then instruct the firefly-swarm to move as per the proposed Firefly-Swarm Next Hop Selection Algorithm (Algorithm 1).
(h) If current node is a midway node, and the route in the routing table is old, then instruct the firefly-swarm to move as per the proposed Firefly-Swarm Next Hop Selection Algorithm (Algorithm 1).
(i) If the broadcast counter exceeds the count of the maximum broadcast counter permissible, then return a route error (RERR), and then Goto START ROUTE DISCOVERY MECHANISM step for current node.

3. Route discovery retried: While waiting for the route reply packet, if the predefined time expires, a timeout error is generated. Next, the routing process has to restart again from Step (1).
4. Initiate Route Reply:

(a) On reaching the destination, the firefly-swarm initiates a route reply packet for the source node.
(b) If current node is midway between the destination and the source, then transmit a route reply using the routing table tuple which was previously updated using lastAddress content of the firefly-swarm during the MOTION OF RECEIVED FIREFLY-SWARM step.

5. Transmission of data packets: The discovered reliable route is used to transfer the data till the entire data has been transferred.

10.4 Performance Evaluation

The proposed EFR protocol is evaluated by simulations using GloMoSim [6] and compared against the DSR protocol in VANET environment using the performance metrics:

- Throughput: Number of bits successfully delivered from source to destination per unit time (in bits/second).
- Packet delivery ratio (PDR): Number of packets received by destination divided by the number of packets generated by the source node.
- Loss packet percentage: Percentage of packets lost divided by the packets generated by the source.
- End-to-end delay: Time taken for a packet to get delivered from source to destination (in milliseconds).

Table 10.1 Simulation parameters

Name	Value
Terrain dimension	$(2 \times 2)\,km^2$
Maximum speed	50 m/s
Total simulation time	15 min
Mobility	Random waypoint
Number of nodes	Up to 70
Range of transmission	376.782 m
Medium access control protocol	IEEE 802.11
Network protocol	IP
Propagation limit	-111.0 decibel
Radio frequency	2.4 GHz

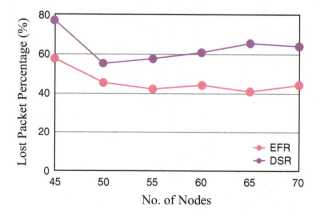

Fig. 10.3 Lost packet percentage vs. no. of nodes

10.4.1 Simulation Parameters

The main simulation parameters are captured in Table 10.1. All other parameters are default GloMoSim parameters.

10.4.2 Simulation Results

First, the node's speed is kept constant at 50 m/s. The number of nodes is varied, and the impact of this variation on the loss packet percentage, PDR, throughput, and end-to-end delay is investigated for the EFR vs. DSR protocols. The results are captured in Figs. 10.3, 10.4, 10.5, and 10.6, respectively.

In Fig. 10.3, it can be observed that the loss packet percentage generated by EFR scheme is significantly lower than that generated by DSR when the number of nodes is increased. This is attributed to the fact that the route search mechanism

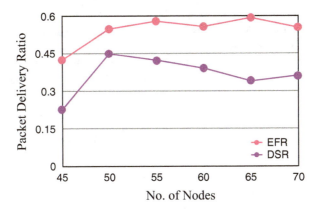

Fig. 10.4 Packet delivery ratio v/s no. of nodes

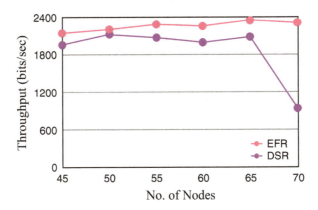

Fig. 10.5 Throughput vs. no. of nodes

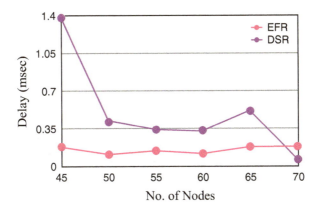

Fig. 10.6 Average delay vs. no. of nodes

implemented in EFR is a more natural approach compared to the DSR route discovery approach, which consists of flooding the control packets.

In Fig. 10.4, it can be observed that the packet delivery ratio generated by EFR is significantly higher than that generated by DSR since EFR provides a more reliable route compared to DSR. In the same trend, in Fig. 10.5, it can be observed that the throughput obtained for EFR is also higher than that obtained for DSR. It is also observed that when the number of nodes exceeds 65, the throughput generated by DSR drops significantly, whereas that generated by EFR is almost stagnant. This is attributed to the fact that the route discovery is faster and more reliable in EFR than in DSR, since in EFR, many fireflies work in parallel to search for destination, which speeds up the connection establishment time.

In Fig. 10.6, it can be observed that the end-to-end delay obtained using EFR is significantly less than that obtained using DSR up to a certain increase in the number of nodes. But in highly dense scenarios, i.e., when the number of nodes is greater than 65, a crossover point occurred after which the end-to-end delay generated by DSR turns out to be better than that generated by EFR. This might be attributed in such situation, a sudden drop has been observed in the throughput generated by EFR, which in turn has led to less number of data packets that have reached the destination quickly. Second, the number of nodes is kept constant at 60. The node's speed is varied, and the impact of this variation on the loss packet percentage, PDR, throughput, and end-to-end delay is investigated for the EFR vs. DSR protocols. The results are captured in Figs. 10.7, 10.8, 10.9, and 10.10, respectively.

In Fig. 10.7, it can be observed that the packet loss obtained using EFR is significantly less than that obtained using DSR when the node's speed is increased. This is attributed to the efficiency of the route discovery process in the EFR scheme compared to its counterpart process in the DSR scheme, in high mobility scenarios. For the same reason, it is observed in Fig. 10.8 that the PDR generated by EFR is significantly higher than that generated by DSR.

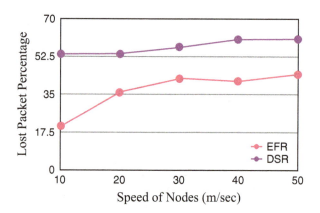

Fig. 10.7 Lost packet percentage vs. speed of nodes

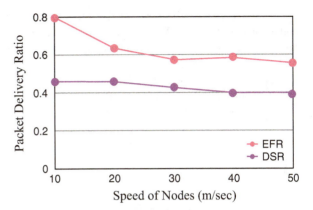

Fig. 10.8 Packet delivery ratio vs. speed of nodes

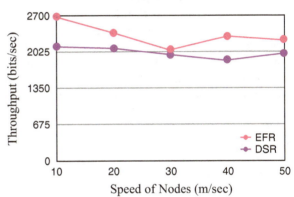

Fig. 10.9 Throughput vs. speed of nodes

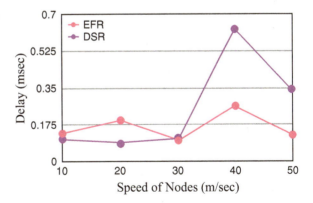

Fig. 10.10 Average delay vs. speed of nodes

In Fig. 10.9, it can be observed that the throughput of EFR is always greater than that of DSR when the node's speed is increased, except when the speed of nodes is 30 m/s, in which case the throughput obtained using EFR is slightly higher than that obtained using DSR. In general, in highly mobile scenarios, EFR outperforms DSR in terms of throughput.

In Fig. 10.10, it can be observed that for an initial speed up to 30 m/s, the end-to-end delay generated by EPR is higher compared to that generated by DSR, but as the node's speed increases to more than that threshold, the end-to-end delay generated by DSR shoots up significantly, whereas that generated by EFR is kept consistent at the same range that it was when the node's speed was less than 30 m/s. This means that in highly mobile scenarios, EPR outperforms DSR in terms of average delay, which is again attributed to its intrinsic fast route discovery process of EFR.

10.5 Conclusion

In this paper, we have proposed an efficient firefly routing protocol with angular routing for VANETs (called EFR), which uses social interaction among fireflies to determine the best route for message passing from source to destination. The proposed EFR protocol is evaluated by simulations, showing that: (1) under varying number of nodes and a fixed node's speed, EFR outperforms DSR in terms of loss packet percentage, PDR, throughput, and end-to-end delay; and (2) under varying node speeds (i.e., high mobility scenario) and a fixed number of nodes, EFR still outperforms DSR in terms of the aforementioned performance metrics. As future work, the EFR technique can be combined with the Q-learning approach to yield a more powerful route search algorithm in terms of enhanced performance. In addition, it is desirable to test the performance of EFR under real VANETs' application scenarios.

Acknowledgements This work is partially supported by a grant from the National Funding from the FCT—Fundação para a Ciência e a Tecnologia through the UID/EEA/500008/2013 Project, by Finep, with resources from Funttel, Grant No. 01.14.0231.00, under the Radiocommunication Reference Center (Centro de Referência em Radiocomunicações-CRR) Project of the National Institute of Telecommunications (Inatel), Brazil, and by the Brazilian National Council for Research and Development (CNPq) via Grant No. 309335/2017-5, all held by the fifth author.

References

1. S. Olariu, M.C. Weigle, *Vehicular Networks: From Theory to Practice*. Chapman & Hall/CRC Computer and Information Science Series (Chapman & Hall/CRC, Boca Raton, 2017), 472 pp. ISBN 9781138116597
2. S.K. Dhurandher, G.V. Singh, Weight based adaptive clustering in wireless ad hoc networks, in *Proceedings of IEEE International Conference on Personal Wireless Communication (ICPWC 2005)*, New Delhi, January 2005, pp. 95–100
3. S. Harrabia, W. Chainbi, K. Ghedira, A multi-agent approach for routing on vehicular ad-hoc networks, *Proceedings of the 4th International Conference on Ambient Systems, Networks and Technologies (ANT 2013)*. Procedia Computer Science, vol. 19 (2013), pp. 578–585
4. S. Bitam, A. Mellouk, S. Zeadally, Bio-inspired routing algorithms survey for vehicular ad hoc networks. IEEE Commun. Surv. Tutor. **17**, 843–867 (2015)
5. X.S. Yang, *Nature-Inspired Metaheuristic Algorithms* (Luniver Press, Bristol, 2008)

6. GlomoSim: Global Mobile Information Systems Simulation Library. https://pcl.cs.ucla.edu/project/glomosim (Last visited Apr. 15, 2018)
7. X.S. Yang, X. He, Firefly algorithm: recent advances and applications. Int. J. Swarm Intell. **1**(1), 36–50 (2013)
8. X.S. Yang, Chaos-enhanced firefly algorithm with automatic parameter tuning. Int. J. Swarm Intell. Res. **2**(4), 1–11 (2011)
9. D.B. Johnson, D.A. Maltz, Dynamic source routing in ad hoc wireless networks, in *Mobile Computing*, Ch. 5, ed. by T. Imielinski, H. Korth (Kluwer Academic Publishing, Norwell, 1996), pp. 153–181
10. R. Brendha, V.S.J. Prakash, A survey on routing protocols for vehicular ad hoc networks, in *Proceedings of the 4th IEEE International Conference on Advanced Computing and Communication Systems (ICACCS)*, Coimbatore, January 6–7, 2017
11. S. Chhabra, R. Kumar, Efficient routing in vehicular ad-hoc networks using firefly optimization, in *Proceedings of the International Conference on Inventive Computation Technologies (ICICT)*, August 26–27, 2016, Coimbatore, pp. 1–7. https://doi.org/10.1109/INVENTIVE.2016.7830212
12. A.M. Oranj, R.M. Alguliev, F. Yusifov, S. Jamali, Routing algorithms for vehicular ad hoc network based on dynamic ant colony optimization. Int. J. Electron. Eng. **4**(1), 79–83 (2016)

Chapter 11
Exploring the Application of Random Sampling in Spectrum Sensing

Hayat Semlali, Najib Boumaaz, Asmaa Maali, Abdallah Soulmani, Abdelilah Ghammaz, and Jean-François Diouris

11.1 Introduction

Software defined radio (SDR) is a multi-mode multi-standard reconfigurable wireless communication system, in which a major part of the treatment is performed in software [1]. The principle of such communication systems presents many advantages but also poses many technological challenges such as the need of high sample rates and very selective filters treating large bands.

The ultimate evolution of software radio is the cognitive radio [2], whose objective is to allow the equipment to choose the best conditions for communications to satisfy the user needs. Cognitive radio aims to solve the problem of saturation of the wireless communication network by an optimization at the terminals. The first enabling stage in cognitive radio is the spectrum sensing operation, where the goal is to obtain the status of the spectrum (free/occupied), so that the spectrum can be accessed by a secondary user without interference with the primary user. The techniques to detect the spectrum have been studied extensively in the literature [3–5]. In this paper we are interested to the energy detector method, which is the most common way of spectrum sensing due to its low computational and

---

H. Semlali (✉) · N. Boumaaz · A. Maali · A. Soulmani · A. Ghammaz
Laboratory of Electrical Systems and Telecommunications, Faculty of Sciences and Technology, Cadi Ayyad University, Marrakech, Morocco

J.-F. Diouris
Department Electronic and Telecommunications Institute of Rennes (IETR – UMR 6164), Polytechnic School of the University of Nantes, Nantes, France

© Springer Nature Switzerland AG 2019
I. Woungang, S. K. Dhurandher (eds.), *2nd International Conference on Wireless Intelligent and Distributed Environment for Communication*, Lecture Notes on Data Engineering and Communications Technologies 27, https://doi.org/10.1007/978-3-030-11437-4_11

Fig. 11.1 Block diagram of a frequency domain energy detector based on the c_k frequency components for the band of interest

implementation complexities and due to the fact that no knowledge on the signals is needed. The block diagram of a frequency domain energy detector is shown in Fig. 11.1.

After the signal is digitized by an analog to digital converter (ADC), the frequency components of the multi-band signal are calculated and we use only the P frequency components of the band of interest. The output is squared and averaged to estimate the received signal energy. The estimated energy, T_{ED}, is then compared with a threshold, λ_E, to decide whether a signal is present (H_1) or not (H_0) [6, 7].

The application of the random sampling in software radio and cognitive radio systems presents several advantages compared to the uniform sampling: There is more flexibility in sampling frequency choices, the constraints on signal filtering operation are reduced and the aliases of the spectrum are reduced (or completely eliminated in the case of a stationary sampling process [8]. Therefore, the constraints on the various elements of the transmission chain are reduced.

In the literature different works based on the energy detector sensing method are presented. In this work, we propose an approach based on random sampling as a sampling mode [9], on the energy detector method [3–6] as the method of spectrum sensing, and on the LU (Gaussian elimination) direct algorithm [10] as a method of spectral components calculation and channel filtering. The performance of the proposed approach will be evaluated and compared to the uniform sampling case.

To complete our theoretical and simulation study, we realize the proposed spectrum sensing approach based on a real FM radio signal and using a random sampling mode. The case study signal is a real FM radio signal captured in the open source GNU-radio environment [11]. The performance of this application will be evaluated in order to demonstrate the feasibility of spectrum sensing operation with a random sampling mode.

11.2 Spectral Components Calculation Using a Random Sampling Mode

A non-uniform sampling process converts a continuous analog signal $x(t)$ into a time discrete representation $x_s(t)$ (Fig. 11.2) where the sampling instants are not uniformly distributed.

In this work, we propose to use a jittered random sampling (JRS) sequence [12], among the most used in the literature.

11 Exploring the Application of Random Sampling in Spectrum Sensing

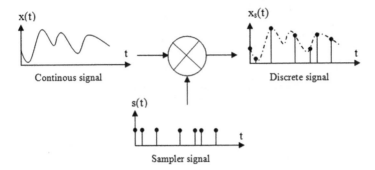

Fig. 11.2 Random sampling principle

The samples are identified by a sample time (t_i) and a value $x_i = x(t_i)$ with i varying from 1 to N. N is the number of samples.

The reconstructed signal is defined by the following expression:

$$\widehat{x}(t) = \sum_{k=1}^{M} c_k \exp(2j\pi f_k t) \tag{11.1}$$

The f_k coefficients are chosen in the signal bandwidth and the c_k frequency components are calculated from the minimization of the squared error defined by:

$$E_q^2 = \|AC - X_s\|^2 \tag{11.2}$$

where

- $X_s^t = [x(t_1), x(t_2) \ldots x(t_N)]$ is a vector of dimension equal to the number of samples N.
- A is a matrix of dimension $N \times M$ formed by the elements of the base $A_m(t_n)$.
- C is a vector of dimension M with the complex elements c_k to calculate.

The minimum of (11.2) is obtained by the solution of the system of linear equations:

$$A^H AC = A^H X_s \tag{11.3}$$

The c_k elements are calculated using the LU (Gaussian elimination) direct algorithm [10].

Once the vector C and its associated frequencies are known, the channel selection can easily be performed. So, if we are interested to a particular band, the channel selection is achieved in two steps: the calculation of all frequency components of the multiband signal followed by the channel selection using only the corresponding c_k coefficients of the band of interest in Eq. (11.1).

11.3 Spectrum Sensing Based on the Energy Detector Method

The energy detection is a technique used to detect the presence of unknown signals [3]. Therefore, it is used in spectrum sensing to detect the occupancy of the radio frequency spectrum without a priori knowledge of these signals. The energy in the band of interest is compared to a detection threshold to decide if this band is occupied or not.

Let us assume that the received signal has the following simple form:

$$x_n = s_n + \omega_n \tag{11.4}$$

where s_n is the signal to be detected, ω_n is an additive white Gaussian noise (AWGN) sample, and n is the sample index. Note that $s_n = 0$ when there is no transmission by primary user. The signal sensed by the CR can be hypothesized as [13, 14]:

$$\begin{cases} H_0 : x_n = \omega_n & \text{(vacant)} \\ H_1 : x_n = s_n + \omega_n & \text{(occupied)} \end{cases} \tag{11.5}$$

H_0 is the hypothesis that the primary user (PU) is not transmitting and hence $s_n = 0$, while H_1 is the hypothesis that the PU is using the channel for transmission. The statistic test for the energy detector can be expressed as:

$$T_{ED} = \frac{1}{P} \sum_{k=1}^{P} |C_k|^2 \tag{11.6}$$

P is the number of the frequency components of the band of interest.

The performance of a detection algorithm can be summarized with two probabilities [3]: P_d the probability of detection and P_{fa} the false alarm probability, which are defined as:

$$\begin{cases} P_{fa} & : \quad \text{Prob} \left\{ T_{ED} > \lambda_E / H_0 \right\} \\ P_d & : \quad \text{Prob} \left\{ T_{ED} > \lambda_E / H_1 \right\} \end{cases} \tag{11.7}$$

P_d is the probability of detecting a signal in the band of interest when it is truly present and P_{fa} is the probability that the test incorrectly decides that the band of interest is occupied when it is not.

To formulate the mathematical equation for the energy detector, the statistic test T_{ED} should be investigated. Under H_0, T_{ED} follows a central chi-square (χ^2) distribution with $2N$ degrees of freedom and under H_1, the statistic test T_{ED} will have a noncentral distribution with the same degrees of freedom and a noncentrality parameter 2SNR [15]. So, we can describe the estimated energy as:

$$T_{ED} \sim \begin{cases} \chi^2_{2N}, & H_0 \\ \chi^2_{2N}\left(2\text{SNR}\right), & H_1 \end{cases} \tag{11.8}$$

Hence, the probability of false alarm P_{fa} and the probability of detection P_d for the ED over AWGN channels are given, respectively, by [15]:

$$P_{fa} = \frac{\Gamma\left(N, \lambda_E/2\right)}{\Gamma(N)} \tag{11.9}$$

$$P_d = Q_N\left(\sqrt{2\text{SNR}}, \sqrt{\lambda_E}\right) \tag{11.10}$$

where λ_E denotes the threshold, $\Gamma(a,x)$ is the incomplete gamma function, $\Gamma(a)$ is the gamma function, $Q_N(a,b)$ is the generalized Marcum Q-function, and SNR denotes the signal to noise ratio which is defined as the ratio of the signal variance σ_s^2 to the noise variance σ_ω^2.

$$\text{SNR} = \frac{\sigma_s^2}{\sigma_\omega^2} \tag{11.11}$$

The performance of a detector can be observed by plotting its receiver operating characteristic curve (ROC curve). The ROC curve is the graphical representation of P_d as a function of P_{fa} for different values of the threshold.

11.4 Results and Discussion

11.4.1 Monte Carlo Simulation for the Proposed Spectrum Sensing Approach

The goal of this section is to evaluate the proposed approach of spectrum sensing operation and compare it with the uniform sampling case. The block diagram of simulations is shown in Fig. 11.3.

After the signal is sampled, we calculate all the frequencies of the multi-band signal using the LU direct algorithm, and we analyze the occupancy of the radio frequency spectrum in the band of interest. Monte Carlo method is used in our application to estimate both the detection and false alarm probabilities for channel occupancy in order to characterize the receiver detection performance.

The case study signal is a group of five carriers separated by 8 MHz, modulated in QPSK, and filtered by a raised cosine filter. Each carrier has a symbol rate $R_s = 4.10^6$ symb/s.

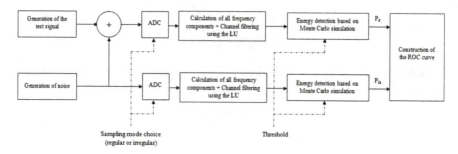

Fig. 11.3 Block diagram of simulation

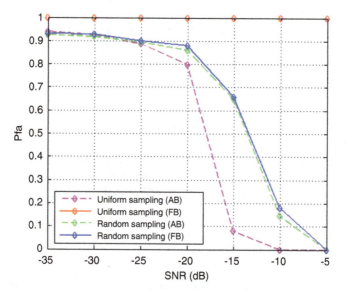

Fig. 11.4 Comparison of the false alarm probability of an energy detector using both sampling modes (random sampling and uniform sampling) for a probability of detection equal to 0.9

To obtain the sampled signal, the input signal is randomly sampled using a JRS process with an average sampling rate $f_s = 1/T_2 = 100$ MHz. The number of samples obtained during the time of observation T is $N = 1000$.

The performance of the proposed approach is evaluated in terms of false alarm probability P_{fa} for different values of the signal to noise ratio, two central frequency values: a center frequency value within the allowed bands (*AB*) and a center frequency value within the forbidden bands (*FB*) and using both sampling modes (random sampling and uniform sampling) in order to demonstrate the utility of the proposed approach (Fig. 11.4). The allowed band is the band of sampling frequencies on which there is no spectrum aliasing [4].

From Fig. 11.4, we can note that in the case of random sampling, the false alarm probability curves are almost similar for the two chosen values of central frequencies. The false alarm probability decreases with increasing SNR.

However, with the use of uniform sampling, we have two cases of false alarm probability curves:

- For central frequency value that is inside the allowed bands, the results are similar to the random sampling case.
- For central frequency value that is inside the forbidden bands, a spectrum aliasing occurs within the channel of interest and hence a great energy is present within this channel even if this channel is free. This explains that the false alarm probability is always equal to 1.

11.4.2 Feasibility Study of the Proposed Approach of Spectrum Sensing Based on a Real FM Radio Signal

In this section, the goal is to demonstrate the feasibility of the spectrum sensing operation using a random sampling mode. In this application, the test study signal is a real FM radio signal with a carrier frequency of 106.5 MHz and a regular oversampling $f_s = 2$ MHz. Figure 11.5 illustrates the block diagram used under GNU radio for capture.

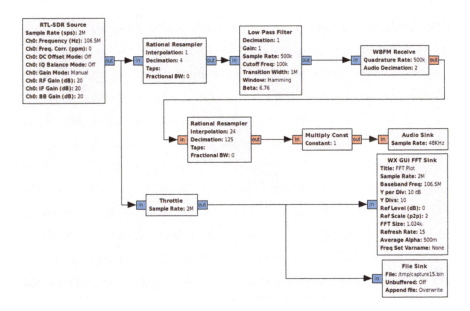

Fig. 11.5 Block diagram used under GNU radio for capture

From this figure, the elements that will allow us to receive the FM band are:

- The RTL2832U key (http://www.realtek.com.tw/products) to receive the coming signal under GNU radio by the block "RTL-SDR source,"
- A *low pass filter* block to limit the frequencies around the listening frequency,
- A WBFM demodulator "WBFM Receive" used for broadcasting,
- A *rational resampler* block to reduce the sampling frequency at a chip frequency,
- And an audio output "Audio sink" in order to listen to the desired station.

The "WX GUI FFT Sink" block is used to observe the spectrum of the listened FM radio signal in real time while the "File Sink" block is used for the capture.

The captured signals are randomly sampled (under Matlab) using a jittered sampling mode with an average frequency $f = 100$ kHz (random decimation by 20).

To realize this application, we made two captures of the signal: a capture when the band is free (only noise is present in the band of interest) and a second shot of the same band when it is occupied (both of the FM radio signal and the noise are present).

The performance of the proposed approach is evaluated in terms of its ROC curves and the false alarm probability for different values of the signal to noise ratio.

Figure 11.6 illustrates the ROC curves and the false alarm probability of the captured real FM radio signal for different values of SNR.

From this figure, we can note that increasing the signal to noise ratio, the signal level becomes higher than the noise level. Therefore, the probability of detection increases and thus the false alarm probability decreases (P_d tends to 1 and P_{fa} tends to 0).

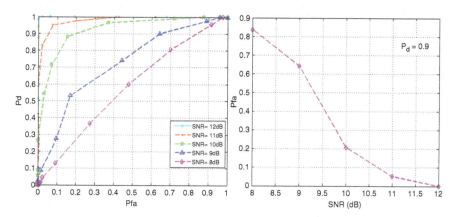

Fig. 11.6 ROC curves and false alarm probability (for a $P_d = 0.9$) of the captured FM radio signal for different values of SNR using a random sampling mode

11.5 Conclusion

In this paper, we were interested in spectrum sensing which is the key function of the cognitive radio. We proposed an approach based on the random sampling as a sampling mode, on the energy detector as a method of spectrum sensing, and on the LU direct algorithm as a method of spectral components calculation and channel filtering. The performance of the proposed approach is evaluated and compared to the uniform sampling case.

To complete the theoretical and the simulation study of our approach, and to demonstrate its feasibility, we implement it on a real FM radio signal using the RTL2832U key. The proposed application is evaluated in terms of the ROC curves and the false alarm probability for different values of SNR. The performances are encouraging so that we can note that random sampling associated with energy detector represents an interesting solution in cognitive radio systems.

References

1. J. Palicot, *De la radio logicielle à la radio intelligente* (Institut Télécom et Lavoisier, Paris, 2010)
2. J. Mitola, G.Q. Maguire, Cognitive radio: making software radios more personal. IEEE Pers. Commun. **6**(4), 13–18 (1999). https://doi.org/10.1109/98.788210
3. T. Yuceka, H. Arsalan, A survey of spectrum sensing algorithms for cognitive radio applications. Commun. Surv. Tutor. **11**(1), 116–130 (2009). https://doi.org/10.1109/surv.2009.090109
4. H. Semlali, N. Boumaaz, A. Soulmani, A. Ghammaz, J.F. Diouris, Energy detection approach for spectrum sensing in cognitive radio systems with the use of random sampling. Wirel. Pers. Commun. **79**(2), 1053–1061 (2014). https://doi.org/10.1007/s11277-014-1917-6
5. L. Claudino, T. Abrão, Spectrum sensing methods for cognitive radio networks: a review. Wirel. Pers. Commun. **95**(4), 5003–5037 (2017). https://doi.org/10.1007/s11277-017-4143-1
6. D. Cabric, A. Tkachenko, R.W. Brodersen, Experimental study of spectrum sensing based on energy detection and network cooperation, in *Proceedings of the 1st International Workshop on Technology and Policy for Accessing Spectrum (TAPAS '06)*, (ACM, New York, August 2006). https://doi.org/10.1145/1234388.1234400
7. D. Cabric, S.M. Mishra, R.W. Brodersen, Implementation issues in spectrum sensing for cognitive radios, in *Asilomar Conference on Signal, Systems and Computers*, November 2004, https://doi.org/10.1109/acssc.2004.1399240
8. I. Bilinskis, A. Mikelsons, *Randomized Signal Processing* (Prentice Hall, Cambridge, 1992)
9. J.J. Wojtiuk, Randomized sampling for radio design, Ph.D. thesis, University of South Australia, School of Electrical and Information Engineering, Australia, 2000
10. A. Bjorck, *Numerical Methods for Least Squares Problems* (SIAM, Philadelphia, PA, 1996)
11. Gnu radio – the free and open software radio ecosystem [Online], Available: http://gnuradio.org/
12. H.S. Shapiro, R.A. Silverman, Alias-free sampling of random noise. SIAM J. Appl. Math. **8**, 225–236 (1960)
13. K.G. Smitha, A.P. Vinod, R. Prashob, Low power DFT filter bank based two-stage spectrum sensing, in *2012 International Conference on Innovations in Information Technology*, 2012, https://doi.org/10.1109/innovations.2012.6207725

14. D. Kakkar, A. Khosla, M. Uddin, Performance evaluation of energy detection in spectrum sensing for cascaded multihop networks over Nakagami-n fading channel. Int. J. Grid Distrib. Comput. **6**(5), 61–70 (2013)
15. F.F. Digham, M.-S. Alouini, M.K. Simon, On the energy detection of unknown signals over fading channels. IEEE Trans. Commun. **55**(1), 21–24 (2007)

Chapter 12
White-Box Cryptography:
A Time-Security Trade-Off
for the SPNbox Family

Federico Cioschi, Nicolò Fornari, and Andrea Visconti

12.1 Introduction

Traditionally, cryptographic primitives are designed to protect data and keys in the black-box attack model, in which the communication end points are trusted, meaning that the cipher execution (encryption/decryption, instantiation with a secret key) cannot be observed or tampered with. However, the assumptions made in the past may often not be applicable in the current technology, such as DRM applications, Pay Tv boxes, and smartphones. For this reason, we refer to the *white-box model* as an attack model in which the adversary has total visibility of the software implementation of the cryptosystem, and full control over its execution platform.

White-Box Cryptography was originally defined [1] as an obfuscation technique intended to implement cryptographic primitives in such a way that an adversary having full access to the implementation and execution platform is unable to extract

---

Some of this work was done as part of the author Federico Cioschi's BSc thesis, Department of Computer Science, Università degli Studi di Milano.

Some of this work was done as part of the author Nicolò Fornari's MSc thesis, Department of Mathematics, University of Trento.

F. Cioschi · A. Visconti (✉)
Department of Computer Science, Università degli Studi di Milano, Milano, Italy
e-mail: federico.cioschi@studenti.unimi.it; andrea.visconti@unimi.it

N. Fornari
The Akkademy, Geneva, Switzerland
e-mail: nicolo.fornari@akka.eu

© Springer Nature Switzerland AG 2019
I. Woungang, S. K. Dhurandher (eds.), *2nd International Conference on Wireless Intelligent and Distributed Environment for Communication*, Lecture Notes on Data Engineering and Communications Technologies 27,
https://doi.org/10.1007/978-3-030-11437-4_12

key information. One may wonder why the adversary should be interested in recovering the key when he/she controls the encryption/decryption software and can use it to encrypt/decrypt data. The reader should be aware that such software might be subject to restrictions as in the case of DRM applications. If a malicious user is able to recover the secret key, then she/he can encrypt or decrypt the data with *any* software on *any* host. In the case of applications enforcing DRM schemes (for example, Sky Go [2], Netflix [3], and Spotify [4]), a key recovery attack would allow an adversary to illegally distribute content to non-subscribers of the service offered by the application. Notice that the definition of white-box cryptography as in [1] is limited to key recovery attempts and does not take into account other attacks such as code lifting, where the attacker attempts to isolate the program code from the implementation environment and directly uses the code itself as a larger key. Therefore, we refer to [5] according to which a white-box implementation of cryptographic primitives does not present any advantage for a computationally bounded adversary in comparison to the adversary dealing with the implementation as a black-box.

White-Box implementations of DES and AES were first proposed by Chow et al. in [6] and [1]. Their approach was to find a representation of the cryptographic algorithm as a network of look-ups in randomized and key-dependent tables. These papers, as well as some others [7, 8], are subjected to algebraic attacks [9–12]. However, such attacks require knowledge of the internal data representation used by the implementation, meaning in practice significant reverse engineering efforts.

A breakthrough from the attacker's side came with the work of Bos et al. [13] who proposed two new attack paradigms which can be automated and do not require reverse engineering efforts. The first attack is known as differential fault analysis (DFA) and can be regarded as the software counterpart of fault-injection attacks on cryptographic hardware. The second one, known as differential computational analysis (DCA), can be thought as a side-channel attack adapted to the white-box attack model.

The DCA attack collects software execution traces for several plaintext encryptions and uses the collected data to perform an analysis similar to the well-known differential power analysis (DPA) to recover the secret key. Since the software traces contain time demarcated physical addresses of memory locations being read and written to, they leak the values of the inputs to the various look-up tables accessed during the white-box encryption operation, which leak enough information to perform the power attack. In [14], Banik et al. further investigate the DCA attack proposing software countermeasures such as randomization of the locations of the look-up tables in memory, in addition to control flow obfuscation. They also develop an attack based on software traces called zero difference enumeration (ZDE). The attack records software traces for several pairs of strategically selected plaintexts and performs a statistical test on the difference of the traces to extract the secret key.

Whereas all published white-box implementations for standard cryptographic algorithms such as DES and AES are prone to practical key extraction attacks as just described, there have been two dedicated design approaches for white-box block ciphers: ASASA by Birykov et al. [15] and SPACE by Bogdanov and Isobe [16]. While ASASA suffers from decomposition attacks, SPACE presents a design for which security against key extraction in the white-box context reduces to the well-studied problem of key recovery for block ciphers in the standard black-box setting. However, SPACE imposes a sometimes prohibitive performance overhead in the real world as it needs many AES calls to encrypt a single block. In [17], Bogdanov et al. address the issue by designing a family of dedicated white-box block ciphers SPNbox and a family of underlying small block ciphers with software efficiency and constant-time execution in mind.

In this chapter, we modify the underlying small block ciphers to increase the number of bits of the key used in each round. This approach can be exploited to make the algorithm faster (we are using the same number of bits of the key reducing the number of rounds) or more secure (we are using the same number rounds increasing the number of bits of the key) than previous.

The remainder of the chapter is organized as follows. In Sect. 12.2, we briefly introduce the white-box implementations of well-known algorithms such as AES. In Sects. 12.3 and 12.4, we describe two families of block ciphers, i.e., the SPACE family and the SPNbox family, designed to be white-box friendly. In Sect. 12.5, we (a) suggest how to speed up the SPNbox family and (b) evaluate the performance of the suggested approach. Finally, discussion and conclusions are drawn in Sect. 12.6.

12.2 White-Box Constructions and Attacks

The idea to avoid key recovery attacks is to mathematically fuse the key with the encryption routine, formally, given a block cipher ϕ, one wants to construct $\psi :$ $\mathbb{F}^n \to \mathbb{F}^n$ such that, fixed a key $\bar{k} \in \mathbb{F}^l$, then $\phi(x, \bar{k}) = \psi(x)$ $\forall x \in \mathbb{F}^n$. Clearly, the secret key k should not be easily retrievable from an attacker knowing ϕ and ψ.

Example Let ϕ, ψ be defined as:

$$\phi(x) := k + x \mod 4 \ x \in \{0, \dots, 3\}$$
$$\psi(x) := S[x] \qquad S = [3, 0, 1, 2]$$

With $k = 3$, we can think of ψ as a trivial white-box implementation of ϕ. In terms of implementation, we can think of ψ as a look-up table.

The first white-box construction for AES was proposed by Chow et al. in [1]. Their work came from a simple idea: given a fixed encryption key and a block cipher, i.e., AES-128, building a look-up table mapping all the possible plaintext to the respective ciphertext is secure against key extraction. Of course for a mapping $\mathbb{F}^{128} \to \mathbb{F}^{128}$ such construction is utterly unfeasible if we use only a huge table. However, many small tables could be used instead.

The approach taken by Chow et al. was to represent a block cipher ϕ as a network of key-dependent look-up tables (see Fig. 12.1 and [1] for details).

Fig. 12.1 Table-based white-box implementation: the key k is scrambled by a network of look-up tables

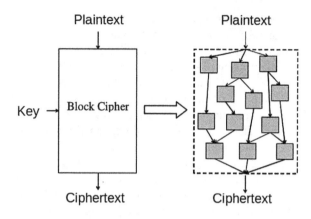

Table 12.1 To the best of our knowledge for each white-box AES implementation available in the literature, there exists an efficient attack

White-box AES impl.	Cryptanalysis	Work factor
Chow [1]	[9]	2^{30}
Karroumi [7]	[10, 12]	2^{22}
Xiao Lai [8]	[10]	2^{32}
Xiao Lai [8] generic linear version	[10]	2^{38}
Xiao Lai [8] affine/non-affine version	[11]	At least 2^{49}

Chow's pioneering work became a reference point for some of the subsequent proposals such as [7] and [8]. For each construction however, an attack was published as summarized in Table 12.1.

Interestingly, the attacks in Table 12.1 require knowledge of the internal data representation used by the implementation. In practice, the level of implementation knowledge required by the attacker is only attainable through significant reverse engineering efforts.

A significant breakthrough from the attacker's perspective became possible by shifting the focus from pure algebraic attacks, as in Table 12.1, to side channel attacks. In [13], Bos et al. describe new approaches to assess the security of white-box implementations. These approaches require *neither* knowledge about the look-up tables used *nor* expensive reverse engineering effort.

Bos et al. introduce a *differential fault analysis* (DFA) attack which is the software counterpart of fault-injection attacks on cryptographic hardware (for more details regarding fault-injection attacks, see [18]). They also describe a *differential computational analysis* (DCA) attack, which is the software counterpart of the differential power analysis for cryptographic hardware (for more details regarding DPA attacks, see [19]).

12   White-Box Cryptography: A Time-Security Trade-Off for the SPNbox Family   157

There are two fundamental concepts that make DFA and DCA attacks so effective:

- Chow's construction is a white-box implementation of AES and, despite being based on look-up tables, the regular structure of AES (SubBytes, Shiftrows, MixColumns, and AddRoundKey) is still preserved.
- Traditional side channel attacks such as fault injection and differential power analysis are extremely more effective in the white-box attack model since the attacker has full control over the execution platform and can thus perform direct measures.

Considering the effectiveness of DCA and DFA attacks, it is time to wonder whether AES is a suitable cipher in the white-box attack model and to consider alternative approaches. Instead of trying to modify well-known ciphers, designed in the black-box model, in order to be resistant in the white-box attack model, it is worth trying to design new ciphers with the white-box attack model in mind. An example of such cipher is SPACE, recalled in Sect. 12.3.

12.3   SPACE

SPACE is a block cipher proposed by Bogdanov and Isobe in [16]. This cipher presents an interesting design for which security against key extraction in the white-box context reduces to the well-studied problem of key recovery for block ciphers in the standard black-box setting.

12.3.1   SPACE Design

SPACE is a generalized Feistel network [20] which encrypts a message $m \in \mathbb{F}^n$ with a secret key $k \in \mathcal{K}$ to a ciphertext $c \in \mathbb{F}^n$. There are three positive integers that we will often use in the description of SPACE: n, n_a, and n_b. In [16], Bogdanov and Isobe fix $n = 128$, $n_a \in \{8, 16, 24, 32\}$, and $n_b = n - n_a$.

The encryption works as follows:

1. Let X^r be the state at round r. The state is seen as l vectors $x_i^r \in \mathbb{F}^{n_a}$:

$$X^r = \{x_0^r, x_1^r, \ldots, x_{l-1}^r\}$$

   where $l = n/n_a$.
2. X^0 is initialized with the value of the plaintext message m.

3. For $r \in \{1, \ldots, R+1\}$, the state updates as follows:

$$X^{r+1} = \left(F_{n_a}^r(x_0^r) \oplus (x_1^r||x_2^r||\ldots||x_{l-1}^r) \right) ||x_0^r$$

where $F_{n_a}^r : \mathbb{F}^{n_a} \to \mathbb{F}^{n_b}$ is the Feistel function and $||$ denotes concatenation.

4. X^{R+1} is the ciphertext c.

The encryption is simple: each round takes x_0^r as input to the Feistel function, then adds $F_{n_a}^r(x_0^r)$ to the rest of the state $(x_1^r||\ldots||x_{l-1}^r)$. The outcome of this operation is saved as the first n_b bits of the new state, while the last n_a are filled by x_0^r.

12.3.2 Feistel Function as a Look-Up Table

Let π_t be a projection, with $t \in \{1, \ldots, 2n\}$, defined as:

$$\pi_t : \quad \mathbb{F}^{2n} \quad \to \quad \mathbb{F}^t$$
$$(x_1, \ldots, x_{2n}) \mapsto (x_1, \ldots, x_t)$$

If we think of $x \in \mathbb{F}^{2n}$ as a vector of bits, we can use π_t to select the t *most* significant bits of x.

Definition 1 Let the Feistel function $F_{n_a}^r$ used by SPACE be defined as:

$$F_{n_a}^r(x) : \mathbb{F}^{n_a} \to \mathbb{F}^{n_b}$$
$$x \mapsto (\pi_{n_b}(\phi_k(\overbrace{0, \ldots, 0}^{n_b}||x))) \oplus r$$

where ϕ_k is a block cipher and r is the round number represented in binary using n_b digits, thus seen as an element of $\mathbb{F}^{n_b}$.

Let us isolate the part of the Feistel function that is round independent.

Definition 2 Let F_{n_a}' be the round independent part of F_{n_a}, defined as:

$$F_{n_a}' : \mathbb{F}^{n_a} \to \mathbb{F}^{n_b}$$
$$x \mapsto \pi_{n_b}(\phi_k(\overbrace{0, \ldots, 0}^{n_b}||x))$$

Now observe that, compared to traditional Feistel networks, SPACE does not use round keys. There is one secret key k used by ϕ_k. Such secret key cannot be hardcoded, hence $\mathbb{F}_{n_a}'$ is implemented as a look-up table. The reader might wonder

12 White-Box Cryptography: A Time-Security Trade-Off for the SPNbox Family

Fig. 12.2 The value of each image of $F'_{n_a}(x)$ is saved as a row in a look-up table. Each row is indexed by the value of x, $x \in \{0, \dots, 2^{n_a} - 1\}$

$$
\overbrace{(0, \dots, 0}^{n_b} \,||\, \overbrace{0, \dots 0, 0)}^{n_a} \mapsto \pi_{n_b}(\phi_k(\overbrace{0, \dots, 0}^{n_b} || 0, \dots 0, 0)
$$
$$
(0, \dots, 0 || 0, \dots 0, 1) \mapsto \pi_{n_b}(\phi_k(0, \dots, 0 || 0, \dots 0, 1))
$$
$$
\vdots
$$
$$
(0, \dots, 0 || 1, \dots 1, 1) \mapsto \pi_{n_b}(\phi_k(0, \dots, 0 || 1, \dots 1, 1))
$$

Table 12.2 Table size for different values of n_a and comparison with other white-box implementations

Cipher	Table size
SPACE-(8,300)	3.84 KB
SPACE-(16,128)	918 KB
SPACE-(24,128)	218 MB
SPACE-(32,128)	51.5 GB
AES (Chow et al.)	752 KB
AES (Xiao Lai)	20.5 MB

We keep the notation SPACE(n_a, R) as in [16], where R is the suggested number of rounds

the reason for designing SPACE over another block cipher ϕ_k when ϕ_k could be implemented as a look-up table directly. It happens that this second option is not possible. If we were to implement ϕ_k as a look-up table, we would need $2^n \cdot n$ bits of space:

$$
\overbrace{(0, \dots 0, 0)}^{n} \mapsto \overbrace{\phi_k(0, \dots 0, 0)}^{n}
$$
$$
(0, \dots 0, 1) \mapsto \phi_k(0, \dots 0, 1)
$$
$$
\vdots
$$
$$
(1, \dots 1, 1) \mapsto \phi_k(1, \dots 1, 1)
$$

Notice that for $n = 128$ such look-up table would be infeasible to construct. The idea of Bogdanov and Isobe (see Fig. 12.2) is to truncate the output of ϕ_k, computed over a smaller domain:

Since the first n_b zeros are used as padding in order to form an n-bit input to provide to ϕ_k, there is no need to store them, hence the look-up table implementation requires $2^{n_a} \cdot n_b$ bits. In Table 12.2, we report the table size of SPACE for $n_a \in \{8, 16, 24, 32\}$.

It is self-evident that not all values of n_a are apt for a real-world application: $n_a = 32$ requires 51.5 GB of storage and even $n_a = 24$, requiring 218 MB, is not suitable for all applications. However, $n_a = 16$ requires only 918 KB, placing itself at the same level of Chow [1] as storage size, see Table 12.2.

12.4 The SPNbox Family

As shown in Sect. 12.3, the SPACE family of space-hard block ciphers [16], using a Feistel structure, offers interesting security properties but it prevents the exploitation of parallel execution. However, as described in [17], using an SPN-type design it is possible to satisfy the requirement of parallelism maintaining a sufficiently high level of space hardness. Therefore, in 2016 Bogdanov et al. described the SPNbox [17] family of space-hard block ciphers, whose structure is shown in Fig. 12.3. Let us explain this approach more formally.

SPNbox-n_{in} is a substitution–permutation network (SPN) with a block length of n bits, a k-bit secret key, and based on n_{in}-bit substitution boxes.

State The state of SPNbox-n_{in} can be represented as a vector of $t = n/n_{in}$ elements of n_{in} bits each:

$$X = \{X_0, \ldots, X_{t-1}\}$$

Key Schedule The k-bit master key is expanded—i.e., $k_0, \ldots, k_{R_{n_{in}}}$ round keys of n_{in} bits—using a key derivation function (KDF)[1]:

$$(k_0, \ldots, k_{R_{n_{in}}}) = KDF(k, n_{in} \cdot (R_{n_{in}} + 1))$$

Round Transformation We encrypt a plaintext X^0 and we get a ciphertext X^R, by applying the following R transformations—e.g., $R = 10$:

$$X^R = (\bigcirc_{r=1}^{R} (\sigma^r \circ \theta \circ \gamma))(X^0)$$

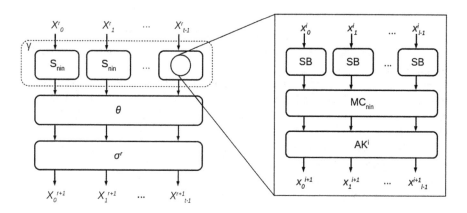

Fig. 12.3 The SPNbox structure with a zoomed view on the inner round

---

[1] For example, PBKDF2 [21–23], ARGON2 [24], Scrypt [25], and so on.

The nonlinear layer γ is a substitution layer in which t key-dependent identical bijective n_{in}-bit S-boxes are applied to the state:

$$\gamma : \mathbb{F}(2^{n_{in}})^t \rightarrow \mathbb{F}(2^{n_{in}})^t$$

$$(X_0, \ldots, X_{t-1}) \mapsto (S_{n_{in}}(X_0), \ldots, S_{n_{in}}(X_{t-1}))$$

These identical S-boxes are realized by an internal small block cipher of block length n_{in} bit.

The linear layer θ, a diffusion layer, applies a $t \times t$ MDS matrix to the state:

$$\theta : \mathbb{F}(2^{n_{in}})^t \rightarrow \mathbb{F}(2^{n_{in}})^t$$

$$(X_0, \ldots, X_{t-1}) \mapsto (X_0, \ldots, X_{t-1}) \cdot M_{n_{in}}$$

The affine layer σ^r adds round-dependent constants to the state:

$$\sigma^r : \mathbb{F}(2^{n_{in}})^t \rightarrow \mathbb{F}(2^{n_{in}})^t$$

$$(X_0, \ldots, X_{t-1}) \mapsto (X_0 \oplus C_0^r, \ldots, X_{t-1} \oplus C_{t-1}^r),$$

with $C_i^r = (r - 1) \cdot t + i + 1$ for $0 \leq i \leq t - 1$.

The Underlying Small Block Ciphers The key-dependent identical n_{in}-bit S-boxes in the γ layer are block ciphers themselves. They are based on the round transformation of AES and consist of $R_{n_{in}}$ rounds operating on a state $x = \{x_0, \ldots, x_{l-1}\}$ of l bytes, where $l = n_{in}/8$ [2]:

$$S_{n_{in}} : \mathbb{F}(2^8)^l \rightarrow \mathbb{F}(2^8)^l$$

$$x \mapsto (\bigcirc_{i=1}^{R_{n_{in}}} (AK^i \circ MC_{n_{in}} \circ SB))(AK^0(x))$$

Notice that: (a) the number of rounds $R_{n_{in}}$ suggested in [17] is $R_{32} = 16$, $R_{24} = 20$, $R_{16} = 32$, and $R_8 = 64$; (b) different matrices are adopted in the $MC_{n_{in}}$ round transformation. More precisely, for $n_{in} = 32$ we use the MC matrix of AES, while in the other cases a sub-matrix of the original one is used. Interestingly, when $n_{in} = 8$, $MC_{n_{in}}$ represents the identity mapping. Notice that, as it happens for the Feistel function in SPACE, in the white-box setting the small block ciphers $S_{n_{in}}$ are implemented as look-up tables.

---

[2] SB, MC, and AK refer to the AES transformations SubBytes, MixColumns, and AddRoundKey, respectively.

12.5 Our Contribution

Three possible issues can be identified in the solution presented in Sect. 12.4.

1. In [17], the authors proposed a black-box implementation of the cipher that uses the AES-NI instructions. During the encryption phase, when $n_{in} = 32$ bits, the A_{32} matrix used by the MixColumns transformation is the same as that used in AES MixColumns. This is true also for $n_{in} = 24$ and $n_{in} = 16$. However, in these cases the A_{32} matrix is not fully used, but two sub-matrices of A_{32}—respectively called A_{24} and A_{16}—are involved in the computation. On the contrary, a different approach has to be adopted in the decryption phase. Let us suppose a scenario in which several clients have to communicate with a server. In this scenario, a black-box implementation of the decryption system may be needed (server side). Such decryption requires the inverse matrices of A_{24} and A_{16} which are not sub-matrices of the inverse of A_{32}. The problem lies on the `aesdec` instruction, provided by AES-NI for decryption, because this instruction is based only on the inverse matrix of A_{32}. In this case, it is not possible to use AES-NI instructions for $n_{in} = 24, 16$.
2. As mentioned in [17], the `aesenc`[3] instruction implies an overhead which depends on the block size (n_{in}). Since increasing parts of the instruction state are not used, it may happen that at halving of n_{in} the overhead doubles. Therefore, the number of `aesenc` instructions needed to encrypt a 128-bit SPNbox state doubles too.
3. Finally, as stated in [17], a possible drawback could be an efficiency bottleneck about the key mixing in the small internal block ciphers. Indeed, the smaller the block size, the more rounds are needed to avoid meet-in-the-middle attacks. This raises the question of how to design an internal block cipher with a faster and secure key mixing.

We try to face these issues by designing an internal block cipher with a faster and secure key mixing. In particular, we modify the inner round (shown in Fig. 12.3) and increase the number of bits of the key used in each round. In doing so, we suggest to replace the classical ShiftRow transformation, omitted in SPNbox [17], with a key-dependent circular bit shift. More precisely, if $n_{in} = 8$, then three bits of the key are required to execute a circular shift on the state (see Fig. 12.4).

This approach allows us to use 11 bits of the key in each round i: eight of them for the AK^i transformation and three for the BitShift. With a larger state, i.e., 16, 24, 32 bits, an independent circular shift is performed on each byte of the state. This means that when $n_{in} = 16, 24, 32$, the number of bits of the key used are 22, 33, and 44, respectively. Although the modified internal structure of AES precludes

---

[3]Provided by AES-NI to make one round of AES encryption.

Fig. 12.4 A BitShift transformation to increase the number of bits of the key used

Fig. 12.5 A modular multiplication to increase the number of bits of the key used

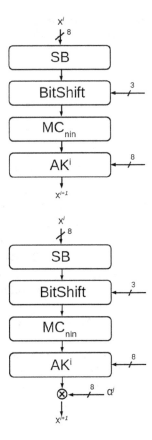

the possibility to make use of the fast AES-NI instructions, the idea suggested can be exploited in two ways:

- If we leave the total number of bits of the key unchanged and increase the number of bits used in each round, thus we are reducing the number of inner rounds, that means speeding up the small internal block cipher;
- If we increase the total number of bits of the key used and leave the number of inner rounds unchanged, thus we are providing the possibility to engineer a block cipher with an improved level of security against meet-in-the-middle attacks.

A further possibility is to add a key-dependent polynomial multiplication at the end of each round. This approach can be implemented with, or without, the BitShift transformation (see Fig. 12.5), allowing the use of the AES-NI instructions. In both cases, the state is multiplied by α^j, where α is a primitive element of $\mathbb{F}(2^{n_{in}})$ and j is an n_{in}-bit exponent provided by the KDF.

Table 12.3 Comparison of the γ layer with and without the BitShift transformation

	γ	γ with BitShift
$n_{in} = 32$, encryption	1.178316 s	0.955048 s
$n_{in} = 32$, decryption	1.447580 s	1.168507 s
$n_{in} = 16$, encryption	3.946748 s	3.222751 s
$n_{in} = 16$, decryption	4.193261 s	3.308678 s
$n_{in} = 8$, encryption	2.547156 s	2.192452 s
$n_{in} = 8$, decryption	2.564750 s	2.250102 s

12.5.1 Performance Evaluation

We compared the performance of the internal layer γ in the black-box setting with and without BitShift transformation for different n_{in} sizes. We run our code on a laptop equipped with Ubuntu 18.04.1 LTS 64 bit, 8 GB RAM, and an Intel® Core™ i3-330M @ 2.13 GHz processor with 3 MB cache. We compile our source code with GCC 7.3.0, -O3 optimization enabled. Table 12.3 shows the time required to encrypt/decrypt one million of different 128-bit plaintexts using the same key.

SPNbox layer γ (see Fig. 12.3) uses 512 key bits (in addition to those needed for the initial AddRoundKey AK^0)—i.e., 512 bit = 16 round × 32 bit ($n_{in} = 32$), or 512 bit = 32 round × 16 bit ($n_{in} = 16$), or 512 bit = 64 round × 8 bit ($n_{in} = 8$). Setting to 512 bit the minimum amount of key bits to be used, our solution will execute: 12 rounds, using 528 key bits ($n_{in} = 32$); 24 rounds, using 528 key bits ($n_{in} = 16$); and finally 47 rounds, using 517 key bits ($n_{in} = 8$).[4]

12.6 Conclusions and Future Works

White-box cryptography aims to ensure the security of cryptographic algorithms in an untrusted environment where an adversary has total visibility of the cryptographic implementations and full control over the software execution platform. In order to make well-known ciphers safe in a white-box context, researchers suggested a number of solutions [1, 6–8]. However, such implementations are subjected to algebraic attacks and side channel attacks, thus researchers developed new ciphers—e.g., SPACE [16] and the SPNbox family [17]—with the white-box attack model in mind.

In this context, our aim is to improve the approach adopted in the SPNbox family [17], focusing on the internal small block ciphers used by Bogdanov et al. In particular, we suggest to increase the number of bits of the key used in each round, replacing the classical ShiftRow operation with a key-dependent circular bit shift,

---

[4]Notice that, when $n_{in} = 32, 16, 8$, the number of key bits used has to be incremented by the appropriate number of key bits used by AK^0.

or introducing a key-dependent polynomial multiplication over the field $\mathbb{F}(2^{n_{in}})$ at the end of the round. The approach suggested can be exploited in two ways: to make the algorithm faster than previous approach, or to make the algorithm more secure against the meet-in-the-middle attack. The testing activities executed on a consumer laptop showed a reduction between 12.27% and 21.10% of the execution time (without AES-NI instructions) of layer γ. Our future work will focus on further increasing the number of bits used by the inner round.

References

1. S. Chow, P. Eisen, H. Johnson, P.C. Van Oorschot, White-box cryptography and an AES implementation, in: International Workshop on Selected Areas in Cryptography (Springer, Berlin, 2002), pp. 250–270
2. Sky Go, http://go.sky.com/. Accessed 13 Nov 2018
3. Netflix, https://www.netflix.com. Accessed 13 Nov 2018
4. Spotify, https://www.spotify.com/. Accessed 13 Nov 2018
5. B. Wyseur, White-Box Cryptography. Ph.D. Thesis, KU Leuven, Department of Mathematics (2009)
6. S. Chow, P. Eisen, H. Johnson, P.C. Van Oorschot, A white-box DES implementation for DRM applications, in ACM Workshop on Digital Rights Management (Springer, Berlin, 2002), pp. 1–15
7. M. Karroumi, Protecting white-box AES with dual ciphers, in International Conference on Information Security and Cryptology (Springer, Berlin, 2010), pp. 278–291
8. Y. Xiao, X. Lai, A secure implementation of white-box AES, in 2nd International Conference on Computer Science and its Applications, 2009, CSA'09 (IEEE, Piscataway, 2009), pp. 1–6
9. O. Billet, H. Gilbert, C. Ech-Chatbi, Cryptanalysis of a white box AES implementation, in International Workshop on Selected Areas in Cryptography (Springer, Berlin, 2004), pp. 227–240
10. Y. De Mulder, P. Roelse, B. Preneel, Cryptanalysis of the Xiao–Lai white-box AES implementation, in International Conference on Selected Areas in Cryptography (Springer, Berlin, 2012), pp. 34–49
11. W. Michiels, P. Gorissen, H.D. Hollmann, Cryptanalysis of a generic class of white-box implementations, in International Workshop on Selected Areas in Cryptography (Springer, Berlin, 2008), pp. 414–428
12. T. Lepoint, M. Rivain, Y. De Mulder, P. Roelse, B. Preneel, Two attacks on a white-box AES implementation, in International Conference on Selected Areas in Cryptography (Springer, Berlin, 2013), pp. 265–285
13. E.A. Bock, J.W. Bos, C. Brzuska, C. Hubain, W. Michiels, C. Mune, E.S. Gonzalez, P. Teuwen, A. Treff, White-box cryptography: don't forget about grey box attacks. Cryptology ePrint Archive, Report 2017/355 (2017)
14. S. Banik, A. Bogdanov, T. Isobe, M. Jepsen, Analysis of software countermeasures for whitebox encryption. IACR Trans. Symmetric Cryptol. **2017**(1), 307–328 (2017)
15. A. Biryukov, C. Bouillaguet, D. Khovratovich, Cryptographic schemes based on the ASASA structure: black-box, white-box, and public-key (extended abstract), in P. Sarkar, T. Iwata (eds.) Advances in Cryptology – ASIACRYPT 2014 (Springer, Berlin, 2014), pp. 63–84
16. A. Bogdanov, T. Isobe, White-box cryptography revisited: space-hard ciphers, in Proceedings of the 22nd ACM SIGSAC conference on computer and communications security (ACM, New York, 2015), pp. 1058–1069

17. A. Bogdanov, T. Isobe, E. Tischhauser, Towards practical whitebox cryptography: optimizing efficiency and space hardness, in International Conference on the Theory and Application of Cryptology and Information Security (Springer, Berlin, 2016), pp. 126–158
18. P. Dusart, G. Letourneux, O. Vivolo, Differential fault analysis on AES, in International Conference on Applied Cryptography and Network Security (Springer, Berlin, 2003), pp. 293–306
19. P. Kocher, J. Jaffe, B. Jun, P. Rohatgi, Introduction to differential power analysis. J. Cryptogr. Eng. **1**(1), 5–27 (2011)
20. H. Feistel, Cryptography and computer privacy. Sci. Am. **228**(5), 15–23 (1973)
21. K. Moriarty, B. Kaliski, A. Rusch, PKCS# 5: Password-Based Cryptography Specification Version 2.1. RFC 8018 (2017)
22. A. Visconti, S. Bossi, H. Ragab, A. Calò, On the weaknesses of PBKDF2, in ed. by M. Reiter, D. Naccache. Cryptology and Network Security (Springer, Berlin, 2015), pp. 119–126
23. A. Visconti, F. Gorla, Exploiting an HMAC-SHA-1 optimization to speed up PBKDF2. IEEE Trans. Dependable Secure Comput. (2018). https://doi.org/10.1109/TDSC.2018.2878697
24. A. Biryukov, D. Dinu, D. Khovratovich, Argon2 (version 1.2). https://password-hashing.net/submissions/specs/Argon-v3.pdf. Accessed 13 Nov 2018
25. C. Percival, S. Josefsson, The scrypt Password-Based Key Derivation Function. RFC 7914 (2016)

Chapter 13
CESIS: Cost-Effective and Self-Regulating Irrigation System

Kaushal A. Shah, Meet Patel, Monil Khasakiya, Saad Kazi, and Pinkesh Khalasi

13.1 Introduction and Motivation

Due to the disturbance in the climate changes, the significant resources to live life in a better manner are getting depleted. A basic resource like water is needed by all the living species. Therefore, we need to start thinking about how to use it wisely and conserve the same for future. We focus on the agriculture aspect where there is a need of a lot of water. As demonstrated in [1, 2], the country like India earns significantly through agriculture. There is a demand to think over using this resource conservatively as in future the wastage of the same will affect the society negatively. If we save water now, then it will be considered as saving money for future.

Water irrigation is a significant aspect of agriculture and it has evolved a lot in recent years [3]. The farmers are looking for various methods that are efficient for water irrigation. The following methods are used widely as discussed in [4]:

- Water irrigation through using buckets and watering cans
- Flood irrigation
- Drip irrigation
- Sprinkler irrigation

These methods require manual intervention and they have following limitations as discussed in [5]:

- Soil nutrients drain off
- Flooding creates erosion

K. A. Shah (✉) · M. Patel · M. Khasakiya · S. Kazi · P. Khalasi
Chotubhai Gopalbhai Patel Institute of Technology, Uka Tarsadia University, Bardoli, India

© Springer Nature Switzerland AG 2019
I. Woungang, S. K. Dhurandher (eds.), *2nd International Conference on Wireless Intelligent and Distributed Environment for Communication*, Lecture Notes on Data Engineering and Communications Technologies 27,
https://doi.org/10.1007/978-3-030-11437-4_13

- Loss of water due to evaporation from plant surfaces
- Water wastage which results in water shortage in areas with less rain fall and production of unhealthy crops

Therefore, we propose *CESIS* to save the water sustainability for the regional area and maintain the crop fields environment friendly by preventing soil and Earth from getting flooded or dried. Moreover, we have to think about the system to save economic cost of water usages for the farmers and for the whole market as discussed in [6–13]. The introduction of IoT devices in the field of agriculture is discussed in [9–11]. The scheduling of irrigation through precision and regulation is discussed in [12, 13]. The low cost factor for automatic irrigation is analyzed in [6, 7]. While several different environmental factors play a key role in agricultural productivity, this chapter focuses purely on the water consumption aspects of agriculture.

The *CESIS* performs the switching mechanism of water pump ON/OFF by using the sensors based on the corresponding threshold level of the moisture from the soil of the field as discussed in [6, 7, 14, 15].

There is a huge demand of food delivery and this inherently requires efficient food production. The climate change is the biggest challenge against developing an effective agriculture system. Therefore, an extra effort must be made for countries producing large amount of revenues through agriculture. As rainfall is decreasing and there is a shortage of reserved water, it results in lack of agriculture resources. As discussed in [16], the area with nonirrigated soil is getting increased day by day due to the reduction of water level on Earth. In addition, as we are not planning the usage of water, it results in wastage of the same.

In any self-regulating irrigation system, as discussed in [6, 7, 14], whenever the dryness level of the soil goes below the threshold, only then the water is supplied. This automation in turn saves a lot of water for us. The irrigation through manually turning the motor ON/OFF is used by the farmers that results in inefficient usage of water, as at times more water than required might get supplied. If less water is supplied or if the supply gets delayed, then the soil dries out and this affects the quality of crops. Therefore, we propose *CESIS* that handles the water supply automatically and reduces water wastage. We also propose a *CESIS* variant that is cheaper as it uses MQTT's publish-subscribe model and Raspberry Pi is not required for this implementation. We have considered only a single-sensor node for the purpose of experimentation of *CESIS*, and data sensed by the sensor is routed to the sink node with the help of MQTT protocol and Wifi module. However, we are working on the multiple-sensor design that forms either a linear or a grid topology as discussed in [17–19].

Organization of the Chapter The rest of the chapter is organized as follows: In Sect. 13.2, the proposed irrigation system is discussed in detail with the help of block diagram, use case diagram, sequence diagram, activity diagram, and class diagram. In Sect. 13.3, we describe the implementation details by looking at all the modules of the proposed irrigation system. We also show the variant to *CESIS* that is more cost effective in this chapter. We summarize our work along with the future directions and scopes in Sect. 13.4.

13.2 The Proposed Irrigation System (*CESIS*)

In this section, we discuss *CESIS* in detail with the help of block diagram, use case diagram, sequence diagram, activity diagram, and class diagram. The variant to *CESIS* does not require Raspberry Pi, and the details of the same are discussed in Sect. 13.3.3.

13.2.1 Block Diagram

In this section, we describe how the components of *CESIS* are arranged, through a schematic diagram known as block diagram. The TE-215—soil moisture sensor senses the moisture level of the soil and sends this information to the NodeMCU. NodeMCU sends moisture data to the Raspberry Pi 3 via MQTT protocol [6, 20, 21]. DHT-11 sensor sends humidity and temperature values to the Raspberry Pi 3. Based on these moisture values, the relay is triggered to turn the water pump ON or OFF. The block diagram of *CESIS* is shown in Fig. 13.1.

13.2.2 Use Case Diagram

In this section, we describe the diagram that shows the behavior of each actor of the *CESIS*. Here, actors are the components of the *CESIS*. The use case diagram of the *CESIS* is shown in Fig. 13.2 that describes all the possible functions that can be performed by the user and the system.

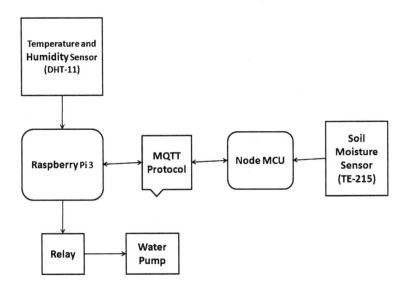

Fig. 13.1 Block diagram

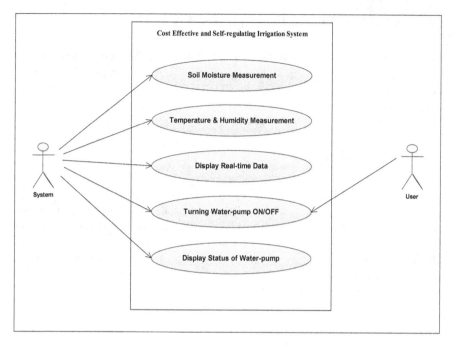

Fig. 13.2 Use case diagram

13.2.3 Sequence Diagram

In this section, we describe the scenarios when a particular event occurs in *CESIS*, through sequence diagram. The sequence diagram shown in Fig. 13.3 describes the flow of moisture data between the TE-215 sensor, NodeMCU, and Raspberry Pi 3. The following sequence of steps are involved in the entire process of *CESIS*:

- The sensor nodes deployed in the farm (we are showing only a single node for the purpose of understanding) pass their sensed data to NodeMCU.
- The NodeMCU is responsible for sending the collected data to Raspberry Pi (there can be more than one NodeMCU based on the topology considered). The NodeMCU uses MQTT protocol for transmitting the data to Raspberry Pi.
- The Raspberry Pi sends the data to the control center.
- The control center has the predefined threshold for the water level of the soil. Based on the value received, the control center takes the decision of whether to turn ON the motor or not. This decision is sent to the Raspberry Pi by the control center.
- Based on the decision received, the Raspberry Pi turns the motor ON or OFF.

13 CESIS: Cost-Effective and Self-Regulating Irrigation System

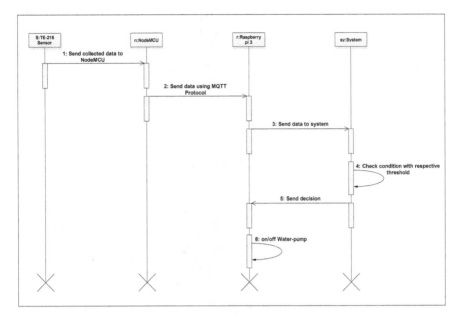

Fig. 13.3 Sequence diagram

13.2.4 Activity Diagram

In this section, we describe the dynamic aspects of *CESIS* with the help of the activity diagram. The activity diagram shown in Fig. 13.4 describes the working flow of the *CESIS*.

13.2.5 Class Diagram

In this section, we describe the relationship and dependencies between the objects of *CESIS*. The class diagram shown in Fig. 13.5 shows various classes with their attributes and functions in the *CESIS*.

The classes and their functionalities of *CESIS* are as follows:

- TE-215 sensor (class) is connected with NodeMCU (class) that has the functions of calculating soil moisture (CalculateSoilMoisture()) and sending (SendSoilMoisture()) the same to Raspberry Pi 3 (class).
- DHT-11 (class) measures the temperature and humidity through GetTemperature() and GetHumidity() methods. It also sends the sensed values to Raspberry Pi 3.

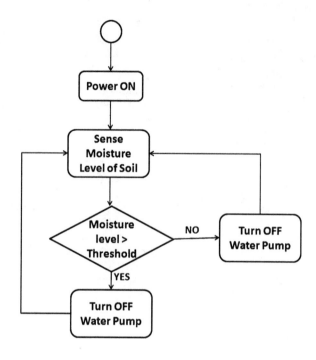

Fig. 13.4 Activity diagram

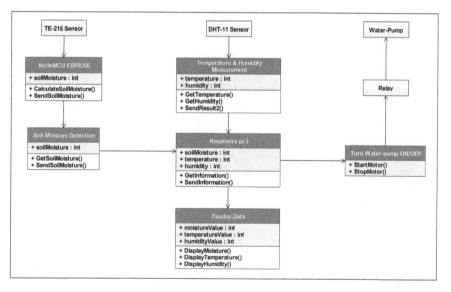

Fig. 13.5 Class diagram

– Raspberry Pi 3 displays the moistureValue, temperatureValue, and humidity-Value with the help of DisplayMoisture(), DisplayTemperature(), and Display-Humidity() methods.
– Raspberry Pi 3 calls the methods StartMotor() and StopMotor() based on the decision to turn the motor ON or OFF.

13.3 Implementation of *CESIS*

In this section, we discuss in detail about the modules that are required to implement *CESIS*.

13.3.1 System Modules

In this section, we describe the system modules required to implement *CESIS*.
Following modules are required to implement the *CESIS*:

1. Soil Moisture Measurement:
   The system measures the amount of moisture in the soil using the TE-215 sensor attached to the NodeMCU. NodeMCU sends these data to Raspberry Pi 3 via MQTT protocol [6, 22–24].
2. Temperature and Humidity Measurement:
   The system measures the amount of temperature and humidity of the atmosphere using the DHT-11 sensor and sends them to the Raspberry Pi 3 [25].
3. Displaying Real-Time Data:
   This system displays real-time statistics of moisture, temperature, and humidity of the soil on web page [26].
4. Turning Water Pump ON/OFF:
   Based on the moisture level in the soil, water pump is turned ON/OFF automatically using the relay. The user can explicitly turn ON/OFF automated or manual irrigation from the web interface [6].
5. Displaying Status and Control of Water Pump:
   The status of the water pump is displayed on the web page. This web page also provides control buttons to turn the water pump ON/OFF explicitly by choosing manual or automated irrigation options [26].

13.3.2 System Model and Implementation

In this section, we describe the system model of *CESIS* and how *CESIS* is implemented.

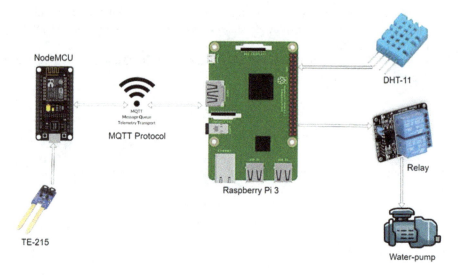

Fig. 13.6 Interconnection between the components (*CESIS*)

The TE-215—soil moisture sensor senses the moisture level of the soil and sends this information to the NodeMCU [22]. The NodeMCU sends the measured soil moisture data to the Raspberry Pi 3 via MQTT protocol [6, 27, 28]. The Raspberry Pi 3 captures these data and detects if the measured soil level is equal to or greater than the required threshold moisture level or not. If the soil moisture is greater than the threshold level, Raspberry Pi 3 triggers high signal to the relay and the water pump is started [27].

The DHT-11 temperature and humidity sensor measures the temperature and humidity of the farm and sends the measured data to Raspberry Pi 3 [6]. The Raspberry Pi 3 hosts a web page using NodeRed and displays real-time data of soil moisture, temperature, and humidity values. Manual and automated irrigation processes can be controlled from the web page [26, 29] (Fig. 13.6).

13.3.3 The CESIS Variant

In this section, we look at the variant of *CESIS* that is more cost effective. The implementation of the same is shown in Fig. 13.7. As can be seen from the figure, MQTT's publish-subscribe model as discussed in [30] is used for automating the process of switching the motor ON/OFF. The dependencies of various components of *CESIS* variant are as follows:

– The sensors send the data of humidity, temperature, and moisture level of soil to the NodeMCU which is working as an MQTT publisher.

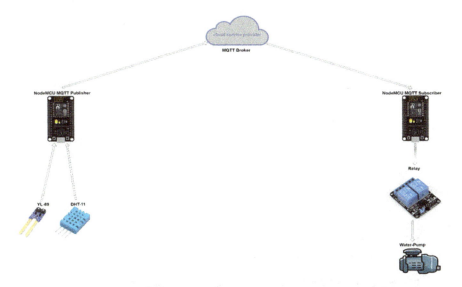

Fig. 13.7 Interconnection between the components (*CESIS* variant)

– The MQTT publisher sends the data to cloud service provider that is working as an MQTT broker.
– The MQTT broker sends the data to the NodeMCU which is working as an MQTT subscriber.
– The MQTT subscriber compares the data with the threshold value and turns the motor ON/OFF accordingly.

13.3.4 Implementation Results

In this section, we describe the implementation results of *CESIS* and also find out the approximate cost of implementing the same. How NodeMCU works as an MQTT publisher and what are the data being sent on the cloud service provider are shown in Fig. 13.8.

The recording of the live data such as humidity and temperature is shown in Fig. 13.9.

The retrieval of data from cloud through NodeMCU working as MQTT subscriber is shown in Fig. 13.10.

We have also created a web page and a mobile application that continuously show the current status of *CESIS*, as shown in Figs. 13.11 and 13.12, respectively. The threshold value for dryness level can be adjusted manually and the switching of motor to ON/OFF value is set according to the threshold value selected.

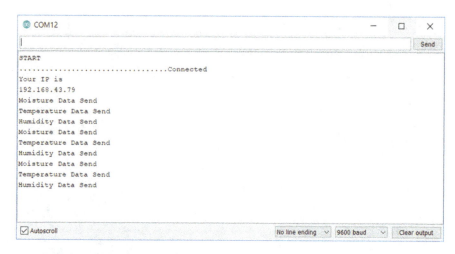

Fig. 13.8 Data sent to cloud service provider

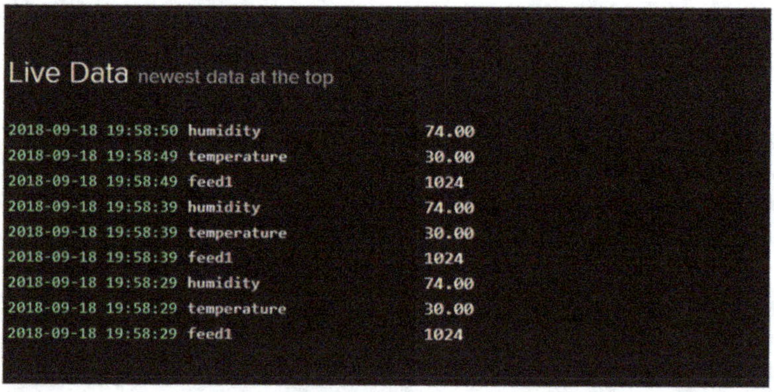

Fig. 13.9 Live data

The implementation cost of *CESIS* variant in Indian Rupee (INR) is shown in Table 13.1.

We have incorporated the assistance through Google assistant [31] that helps the farmers to just speak the commands and accordingly actions would be triggered. We demonstrate how Google assistant actually works in Fig. 13.13.

13.4 Conclusions and Future Scope

This section describes the summary of our work along with some open problems as future work.

13   CESIS: Cost-Effective and Self-Regulating Irrigation System   177

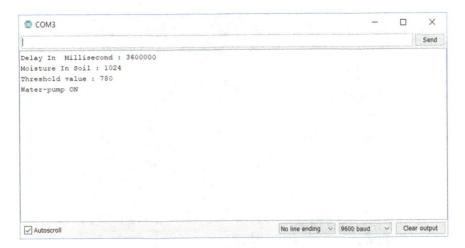

Fig. 13.10 Data retrieval from cloud

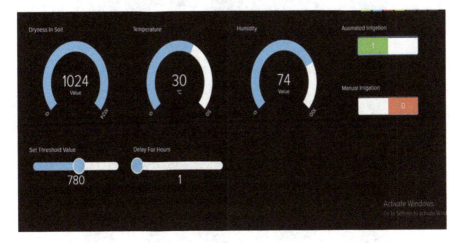

Fig. 13.11 Web page

13.4.1   Conclusions

The *CESIS* is an automated system requiring lesser manual interventions and thus provides various advantages. Water is supplied through *CESIS* only when it is actually required by the soil, i.e., when the dryness value of the soil reaches a threshold. The water is applied directly to the soil by maintaining a moisture ratio to conserve the water. The *CESIS* is efficient and compatible to changing environment as the values of threshold and water supply can be changed according to the need.

Fig. 13.12 Mobile application

The *CESIS* helps farmers and gardeners in watering the plants, by sensing the soil moisture and switching the water pump ON/OFF automatically. The *CESIS* is implemented at relatively low cost. We can install the *CESIS* in home environment as well with lesser efforts and resources. The *CESIS* variant costs only 925 INR that is really cost effective as compared to the other automated irrigation systems. We have provided the support of Google assistant as well.

13 CESIS: Cost-Effective and Self-Regulating Irrigation System

Table 13.1 Implementation cost of *CESIS* variant

Hardware	Quantity	Cost
NodeMCU	2	600 INR
DHT-11	1	100 INR
YE-69	1	150 INR
Channel relay	1	75 INR
Total		925 INR

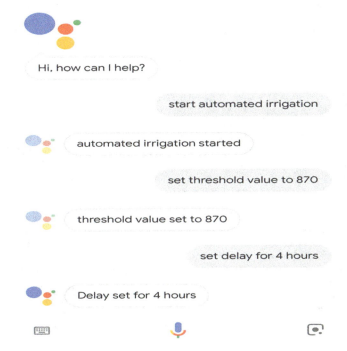

Fig. 13.13 Google assistant support

13.4.2 Future Scope

The *CESIS* works on the value received through the sensors deployed in the field. The amount of water supply always depends on the kind of crops. If the field is carrying RICE crops, then it is not possible to use the current implementation of *CESIS* to deal with the same as values cannot change automatically. However, one can manually adjust the values. Therefore, this customized irrigation system needs to be developed for the desired field with its corresponding water demand as discussed in [32].

We are also working on storing the measured values of temperature and humidity on cloud, for accurately analyzing the soil water requirements.

References

1. S.S. Gill, I. Chana, R. Buyya, IoT based agriculture as a cloud and big data service: the beginning of digital India. J. Organ. End User Comput. **29**(4), 1–23 (2017)
2. A. Patil, M. Beldar, A. Naik, S. Deshpande, Smart farming using Arduino and data mining, in 3rd International Conference on Computing for Sustainable Global Development (INDIACom) (IEEE, Piscataway, 2016), pp. 1913–1917
3. J.T. Srinivasan, V.R. Reddy, Impact of irrigation water quality on human health: a case study in India. Ecol. Econ. **68**(11), 2800–2807 (2009)
4. R. Sivanappan, Prospects of micro-irrigation in India. Irrig. Drain. Syst. **8**(1), 49–58 (1994)
5. R.M. Saleth, Water rights and entitlements in India, in Indian Water Policy at the Crossroads: Resources, Technology and Reforms (Springer, Berlin, 2016), pp. 179–207
6. R.K. Kodali, B.S. Sarjerao, A low cost smart irrigation system using MQTT protocol, in Region 10 Symposium (TENSYMP) (IEEE, Piscataway, 2017), pp. 1–5
7. M. Saleh, I.H. Elhajj, D. Asmar, I. Bashour, S. Kidess, Experimental evaluation of low-cost resistive soil moisture sensors, in International Multidisciplinary Conference on Engineering Technology (IMCET) (IEEE, Piscataway, 2016), pp. 179–184
8. V.H. Andaluz, A.Y. Tovar, K.D. Bedón, J.S. Ortiz, E. Pruna, Automatic control of drip irrigation on hydroponic agriculture: Daniela tomato production, in International Conference on Automatica (ICA-ACCA). (IEEE, Piscataway, 2016), pp. 1–6
9. A. Kamilaris, F. Gao, F. X. Prenafeta-Boldú, M.I. Ali, Agri-IoT: a semantic framework for internet of things-enabled smart farming applications, in 3rd World Forum on Internet of Things (WF-IoT) (IEEE, Piscataway, 2016), pp. 442–447
10. N.K. Verma, A. Usman, Internet of things (IoT): a relief for Indian farmers, in Global Humanitarian Technology Conference (GHTC) (IEEE, Piscataway, 2016), pp. 831–835
11. J. Shenoy, Y. Pingle, IoT in agriculture, in 3rd International Conference on Computing for Sustainable Global Development (INDIACom) (IEEE, Piscataway, 2016), pp. 1456–1458
12. V.V. Hari Ram, H. Vishal, S. Dhanalakshmi, P.M. Vidya, Regulation of water in agriculture field using internet of things, in Technological innovation in ICT for agriculture and rural development (TIAR) (IEEE, Piscataway, 2015), pp. 112–115
13. A.N. Harun, M.R.M. Kassim, I. Mat, S.S. Ramli, Precision irrigation using wireless sensor network, in International Conference on Smart Sensors and Application (ICSSA) (IEEE, Piscataway, 2015), pp. 71–75
14. K. Lekjaroen, R. Ponganantayotin, A. Charoenrat, S. Funilkul, U. Supasitthimethee, T. Triyason, IoT planting: watering system using mobile application for the elderly, in International Computer Science and Engineering Conference (ICSEC) (IEEE, Piscataway, 2016), pp. 1–6
15. N. Ishak, A.H. Awang, N. Bahri, A. Zaimi, GSM activated watering system prototype, in International Conference on RF and Microwave (RFM) (IEEE, Piscataway, 2015), pp. 252–256
16. E. Zaveri, D.S. Grogan, K. Fisher-Vanden, S. Frolking, R.B. Lammers, D.H. Wrenn, A. Prusevich, R.E. Nicholas, Invisible water, visible impact: groundwater use and Indian agriculture under climate change. Environ. Res. Lett. **11**(8), 084005 (2016)
17. K. Shah, D.C. Jinwala, A secure expansive aggregation in wireless sensor networks for linear infrastructure, in Region 10 Symposium (TENSYMP) (IEEE, Piscataway, 2016), pp. 207–212
18. K.A. Shah, D.C. Jinwala, Novel approach for pre-distributing keys in WSNs for linear infrastructure. Wirel. Pers. Commun. **95**(4), 3905–3921 (2017)
19. K.A. Shah, D.C. Jinwala, Privacy preserving, verifiable and resilient data aggregation in grid-based networks. Comput. J. **61**(4), 614–628 (2018)
20. M.F. Leroux, G.V. Raghavan, Design of an automated irrigation system. McGill University Canada, research paper (2005)
21. N. Agrawal, S. Singhal, Smart drip irrigation system using Raspberry Pi and Arduino, in International Conference on Computing, Communication & Automation (ICCCA) (IEEE, Piscataway, 2015), pp. 928–932

22. F. binti Abdullah, N.K. Madzhi, F.A. Ismail, Comparative investigation of soil moisture sensors material using three soil types, in 3rd International Conference on Smart Instrumentation, Measurement and Applications (ICSIMA) (IEEE, Piscataway, 2015), pp. 1–6

23. S. Suradhaniwar, S.A. Sawant, M. Badnakhe, S.S. Durbha, J. Adinarayana, An interoperable wireless sensor network platform for spatio-temporal soil moisture and soil temperature estimation, in Fifth International Conference on Agro-Geoinformatics (Agro-Geoinformatics) (IEEE, Piscataway, 2016), pp. 1–6

24. P. Singh, S. Saikia, Arduino-based smart irrigation using water flow sensor, soil moisture sensor, temperature sensor and ESP8266 WiFi module, in *Region 10 Humanitarian Technology Conference (R10-HTC)* (IEEE, Piscataway, 2016), pp. 1–4

25. R. Shete, S. Agrawal, IoT based urban climate monitoring using Raspberry Pi, in International Conference on Communication and Signal Processing (ICCSP) (IEEE, Piscataway, 2016), pp. 2008–2012

26. P.H. Tarange, R.G. Mevekari, P.A. Shinde, Web based automatic irrigation system using wireless sensor network and embedded Linux board, in International Conference on Circuit, Power and Computing Technologies (ICCPCT) (IEEE, Piscataway, 2015), pp. 1–5

27. M.A. Triawan, H. Hindersah, D. Yolanda, F. Hadiatna, Internet of things using publish and subscribe method cloud-based application to NFT-based hydroponic system, in International Conference on Frontiers of Information Technology (FIT) (IEEE, Piscataway, 2016), pp. 98–104

28. S.S. Solapure, H. Kenchannavar, Internet of things: a survey related to various recent architectures and platforms available, in International Conference on Advances in Computing, Communications and Informatics (ICACCI). (IEEE, Piscataway, 2016), pp. 2296–2301

29. D.B. Ware, D.S. Ware, Irrigation controller with embedded web server. Mar. 7 2006, US Patent 7,010,396

30. U. Hunkeler, H.L. Truong, A. Stanford-Clark, MQTT-S—a publish/subscribe protocol for wireless sensor networks, in 3rd International Conference on Communication Systems Software and Middleware and Workshops (IEEE, Piscataway, 2008), pp. 791–798

31. G. López, L. Quesada, L.A. Guerrero, Alexa vs. Siri vs. Cortana vs. Google assistant: a comparison of speech-based natural user interfaces. in International Conference on Applied Human Factors and Ergonomics (Springer, Berlin, 2017), pp. 241–250

32. A. Sonit, K. Hemlata, J. Sinha, P. Katre, Optimization of water use in summer rice through drip irrigation. J. Soil Water Conserv. **14**(2), 157–159 (2015)

Chapter 14
Maximum Eigenvalue Based Detection Using Jittered Random Sampling

Asmaa Maali, Sara Laafar, Hayat Semlali, Najib Boumaaz, and Abdallah Soulmani

14.1 Introduction

The advent of new wireless services increases the need of spectrum for wireless communication. This increase leads to spectrum penury. Cognitive radio (CR) technology has been proposed as an optimistic solution to optimize the spectrum resources and to solve the problem of scarcity.

In cognitive radio networks, the primary users (PU) can be defined as the users who have higher priority in the usage of a specific frequency band. The secondary users (SU) are the users who have lower priority [1]. SU use the frequency band in such way they do not interfere with the primary users. In this context, spectrum sensing (SS) which is the basic function in cognitive radio systems allows to detect the spectrum occupancy. Several techniques of spectrum sensing are presented in the literature including matched filter detection (MF) [2, 3], energy detection (ED) [4, 5], and cyclostationary feature detection (CSD) [2, 6]. Each one has its own pros as well as cons. The matched filter detection provides optimal detection but requires a complete knowledge of the PU signal. The cyclostationary feature detection is based on exploiting cyclostationary features of the received signal; it offers good performance for detection, but requires a partial knowledge of the PU characteristics and high computation time to complete sensing. The energy detection is a major and basic method due to its ease implementation and low complexity. Unlike other methods, energy detection does not need any prior information of PU's signal. On the negative side, its performance at low SNR is not satisfactory. The maximum

---

A. Maali (✉) · S. Laafar · H. Semlali · N. Boumaaz · A. Soulmani
Department of Physics, Laboratory of Electrical Systems and Telecommunications,
Faculty of Sciences and Technology, Cadi Ayyad University, Marrakech, Morocco

© Springer Nature Switzerland AG 2019
I. Woungang, S. K. Dhurandher (eds.), *2nd International Conference on Wireless Intelligent and Distributed Environment for Communication*, Lecture Notes on Data Engineering and Communications Technologies 27,
https://doi.org/10.1007/978-3-030-11437-4_14

eigenvalue detection (MED) is a method that is widely studied in recent research topics [7–12]. This technique uses the covariance matrix theory (CMT) for sensing the primary user.

As we noticed above, one of the key functions of cognitive radio is SS that is the most important skill to establish cognitive radio networks. It aims to determine the availability of the spectrum and to detect the presence of the primary user when a user operates in a licensed band. The SS in cognitive radio system requires high sampling rate, high-resolution analog to digital converters (ADC) with large dynamic range, and high-speed signal processors. Therefore, it is difficult to implement this function if the hardware cannot satisfy these criteria.

The application of random sampling (RS) in cognitive radio systems seems to present several advantages compared to the uniform sampling case [13]: there is more flexibility in sampling frequencies, the constraints on signal filtering operation are reduced, and the aliases of the spectrum are attenuated (or completely eliminated in the case of a stationary sampling sequence) ([14], p. 36). Therefore, the constraints on the various elements of the transmission chain are reduced.

In the literature, many works on the maximum eigenvalue detector method were presented based on uniform sampling. The purpose of this work is to use the random sampling in the maximum eigenvalue technique. The combination of the random sampling with MED will help to optimize cognitive radio systems, as it will take advantage of RS benefits cited above.

The remainder of this paper is structured as follows: Sect. 14.2 reviews the maximum eigenvalue method. Section 14.3 presents the random sampling mode. In Sect. 14.4, we discuss the obtained results of the proposed approach. Summary of this work and the conclusions are presented in Sect. 14.5.

14.2 Maximum Eigenvalue Detector Basis

In this work, we focus on the crucial function in CR which is spectrum sensing. It allows to sense the PU transmissions in a given frequency band at a given time and then to make decision about this band.

Suppose that the received signal has the following simple form:

$$x_n = s_n + \omega_n \tag{14.1}$$

where s_n denotes the primary user signal (signal to be detected) and ω_n is an additive white Gaussian noise (AWGN) and n is the sample index. We note that $s_n = 0$ when there is no transmission by the primary user.

The signal detection problem is equivalent to the difference between the following states [15]:

$$\begin{aligned} H_0 &: x_n = \omega_n \text{ (vacant)} \\ H_1 &: x_n = s_n + \omega_n \text{ (occupied)} \end{aligned} \tag{14.2}$$

14 Maximum Eigenvalue Based Detection Using Jittered Random Sampling

H_0 represents the null hypothesis that the primary user is absent; while H_1 states that a primary user is present in the channel of the interest.

The performance of any spectrum sensing technique can be indicated by the probability of false alarm (P_{FA}) and the probability of detection (P_D) which are defined as follows [4]:

$$
\begin{aligned}
P_{FA} &: \mathrm{Prob} \left\{ T > \lambda / H_0 \right\} \\
P_D &: \mathrm{Prob} \left\{ T > \lambda / H_1 \right\}
\end{aligned}
\tag{14.3}
$$

Maximum eigenvalue based detection is a method widely developed. It can be considered as the most reliable one among the presented methods. It presents many advantages: it does not require any knowledge of signal properties [7–12], allows a good detection at low signal to noise ratio (SNR). In the maximum eigenvalue detection technique, random matrix theory (RMT) is used to formulate the detection algorithm depending on sample covariance matrix of the received signal.

Let L be the number of consecutive samples, $\widehat{x}(n)$ an estimation of the received signal, $\widehat{s}(n)$ an estimation of primary signal to be detected and $\widehat{w}(n)$ an estimation of the noise. We define the following vectors form:

$$
\widehat{x}(n) = [x(n), x(n+1), \ldots \ldots \ldots x(n+L-1)]^T
\tag{14.4}
$$

$$
\widehat{s}(n) = [s(n), s(n+1), \ldots \ldots \ldots s(n+L-1)]^T
\tag{14.5}
$$

$$
\widehat{w}(n) = [w(n), w(n+1), \ldots \ldots \ldots w(n+L-1)]^T
\tag{14.6}
$$

The approximated statistical covariance matrix $\widehat{R}_x$ is defined by [10] as:

$$
\widehat{R}_x(N_s) =
\begin{bmatrix}
\xi(0) \, \xi(1) & \ldots & \xi(L-1) \\
\xi(1) \, \xi(0) & \ldots & \xi(L-2) \\
\lambda \begin{pmatrix} \vdots \end{pmatrix} \lambda \begin{pmatrix} \vdots \end{pmatrix} & \lambda \begin{pmatrix} \vdots \end{pmatrix} & \lambda \begin{pmatrix} L:2 \end{pmatrix} \\
\xi(L-1) \, \xi(L-2) & \ldots & \xi(0)
\end{bmatrix}
\tag{14.7}
$$

where $\xi(l)$ is the sample auto-correlations of the received signal. It is described as:

$$
\xi(l) = \frac{1}{N_{MED}} \sum_{m=0}^{N_{MED}-1} x(m)x(m-l), \, l = 0, 1, \ldots, L-1
\tag{14.8}
$$

with N_{MED} the number of available samples.

Based on the random matrix theory (RMT), the probability of false alarm for maximum eigenvalue detection is given as:

Table 14.1 Numerical table

t	−3.90	−3.18	−2.78	−1.91	−1.27	−0.59	0.45	0.98	2.02
$F_1(t)$	0.01	0.05	0.10	0.30	0.50	0.70	0.90	0.95	0.99

$$P_{FA} \approx 1 - F_1 \left(\frac{\lambda_{MED} N_{MED} - \mu}{\vartheta} \right) \tag{14.9}$$

$$\mu = \left(\sqrt{\left(N_{MED} - 1 \right)} + \sqrt{L} \right) 2 \tag{14.10}$$

$$\vartheta = \left(\sqrt{(N_{MED} - 1)} + \sqrt{L} \right) \left(\frac{1}{\sqrt{(N_{MED} - 1)}} + \frac{1}{\sqrt{L}} \right)^{1/3} \tag{14.11}$$

F_1 is the cumulative distribution function of the Tracy-Widom distribution of order 1. To compute the F_1^{-1} at certain points, Table 14.1 can be used.

From a given P_{FA}, N_{MED}, and L, the sensing threshold used for the decision process is given in expression (14.12).

$$\lambda_{MED} = \frac{\left(\sqrt{N_{MED}} + \sqrt{L} \right)^2}{N_{MED}} \left(1 + \frac{\left(\sqrt{N_{MED}} + \sqrt{L} \right)^{-\frac{2}{3}}}{(N_{MED})^{\frac{1}{6}}} F_1^{-1} (1 - P_{FA}) \right) \tag{14.12}$$

The detection algorithm of the maximum eigenvalue detection is summarized as follows:

Step 1: Compute the sample auto-correlations in (14.8) and form the sample covariance matrix defined in (14.7).

Step 2: Find the maximum eigenvalue ξ_{max} of the sample covariance matrix by using the eigenvalue decomposition techniques.

Step 3: Decide: if $\xi_{max} > \lambda_{MED} * \sigma_\omega^2$, then the primary user exists, otherwise, it does not.

14.3 Random Sampling Mode

Random sampling process consists in converting a continuous analog signal $x(t)$ into a discrete time representation $x_s(t)$ (Fig. 14.1) where the sampling instants are not uniformly distributed.

The application of the random sampling makes it possible to reduce the aliases of spectrum (or to eliminate them in the case of a stationary sequence), thus helping to reduce the constraints on the various elements of the transmission chain [13, 16]. In

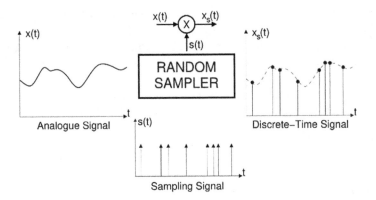

Fig. 14.1 Sampling random principle

Fig. 14.2 Simulation diagram

the literature, there are two most commonly used random sampling modes, namely ARS (additive random sampling) and JRS (jittered random sampling). In this work, we propose to use JRS mode for its ease implementation.

The jittered random sampling (JRS) is a random process where the sampling times are described by the following expression:

$$t_n = nT + \tau_n \quad T > 0, n = 1, 2, \ldots \ldots \quad (14.13)$$

T represents the mean period of sampling.

τ denotes a set of independent random variables identically distributed with a probability density $p(\tau)$, a variance σ^2, and mean $= 0$.

14.4 Applications and Simulation Results

In this contribution, our objective is to exhibit the feasibility of the maximum eigenvalue detector applying jittered random sampling as mode of random sampling. In this section, we present the performance of the MED associated with the random sampling.

The simulation diagram is shown in Fig. 14.2.

The test signal is a multi-band signal consisting of five carriers spaced by 80 Hz, modulated with QPSK then filtered by a raised cosine filter with a rounding coefficient (roll-off) of 0.5. Each carrier has a symbol rate of $R = 40$ sym/s. The values considered are suitable for our calculator power.

After digitizing the generated signal, its frequency components are calculated. Then, the band of interest is selected by applying the SVD algorithm. The occupancy of the radio frequency spectrum is analyzed using the spectrum sensing studied method. In this application, we consider an AWGN (additive white Gaussian Noise) channel.

The performance of the studied technique is evaluated in terms of the ROC curve which is the graphical representation of the detection probability P_D versus the false alarm probability P_{FA} for different threshold values and the detection probability for different values of SNR and smoothing factor L.

Figure 14.3 illustrates the ROC curve for the maximum eigenvalue detection method for an SNR $= -19$ db and a smoothing factor $L = 8$. From this figure, we can note that the MED method presents a good detection.

In Fig. 14.4, we evaluate the detection probability of the maximum eigenvalue detector as a function of the number of samples in order to analyze its effect. This figure shows that the detection probability of the MED method increases by increasing the number of samples and the detection probability is good whatever the value of the SNR.

Figure 14.5 shows the detection probability as a function of the number of the smoothing factor L. The probability of detection of the MED increases by increasing the smoothing factor and it is not very sensitive to the smoothing factor for $L \geq 8$.

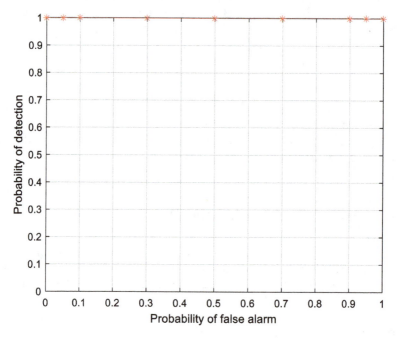

Fig. 14.3 ROC curve (P_D vs. P_{FA}) for SNR $= -19$ dB, $L = 810{,}000$ Monte-Carlo realizations

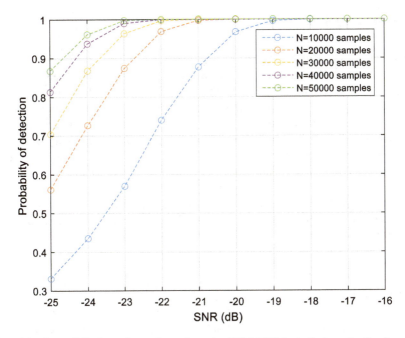

Fig. 14.4 PD vs. SNR, for various values of samples N (10,000 Monte-Carlo realizations)

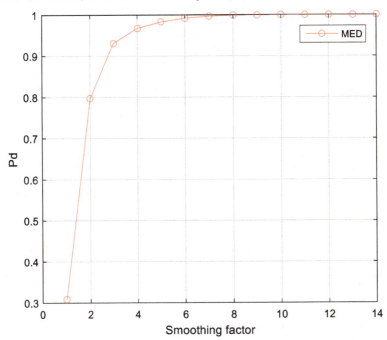

Fig. 14.5 P_D vs. the smoothing factor L, for $P_{FA} = 0.1$ and SNR $= -19$ dB, using 10,000 Monte-Carlo realizations

14.5 Conclusion

For the last several years, cognitive radio has attracted considerable attention in research communities. In this work, we were interested by spectrum sensing which is the crucial function in cognitive radio. We presented the feasibility of using random sampling in the CR system; we combined random sampling with the maximum eigenvalue based detection. The proposed approach is evaluated in terms of its receiver operating characteristics (ROC curve) and its detection probability for different values of signal to noise ratio (SNR) and of smoothing factor L. From the results obtained, we can conclude that we can associate the random sampling and the spectrum sensing technique to be used as a solution in cognitive radio systems. In the future work, we will try to implement this approach for the real scenarios, and to study the performance of this approach in terms of the flexibility in sampling frequencies.

References

1. A. Maali, H. Semlali, N. Boumaaz, A. Soulmani, Energy Detection Versus Maximum Eigenvalue Based Detection: A Comparative Study, in *2017 14th International Multi-Conference on Systems, Signals & Devices (SSD)*, (Marrakech, 2017), pp. 1–4
2. A. Sahai, D. Cabric, Spectrum sensing: fundamental limits and practical challenges, in *IEEE International Symposium on New Frontiers in Dynamic Spectrum Access Networks (DySPAN '05)*, Baltimore, MD, USA, November 2005
3. R. Tandra, A. Sahai, Fundamental limits on detection in low SNR under noise uncertainty, in *Proceedings of the International Conference on Wireless Networks, Communications and Mobile Computing (WirelessCom '05)*, vol. 1, (Maui, Hawaii, USA, 2005), pp. 464–469
4. T. Yucek, H. Arslan, A survey of spectrum sensing algorithms for cognitive radio applications. IEEE Commun. Surv. Tutor. **11**(1), 116–130 (2009). First Quarter
5. H. Urkowitz, Energy detection of unkown deterministic signals. Proc. IEEE **55**(4), 523–531 (1967)
6. Z. Khalaf, A. Nafkha, J. Palicot, Low Complexity Enhanced Hybrid Spectrum Sensing Architectures For Cognitive Radio Equipment. Int. J. Adv. Telecommun. **3**(3&4), 215–227 (2010)
7. Z. Li, H. Wang, J. Kuang, A two spectrum sensing scheme for cognitive radio networks, in *International Conference on Information Science and Technology*, 26–28 March 2011
8. Y. Zeng, Y.C. Liang, Eigenvalue-based spectrum sensing algorithms for Cognitive Radio. IEEE Trans. Commun. **57**(6), 1784–1793 (2009)
9. Y. Zeng, Y.C. Liang, Spectrum-Sensing Algorithms for Cognitive Radio based on Statistical Covariances. IEEE Trans. Veh. Technol. **4**, 58 (2009)
10. Y. Zeng, C. Koh, Y.C. Liang, Maximum eigenvalue detection: theory and application, in *Proceedings of the IEEE ICC*, May 2008, pp. 4160–4164
11. S.K. Sharma, S. Chatzinotas, B. Ottersten, Maximum eigenvalue detection for spectrum sensing under correlated noise, in *IEEE International Conference on Acoustic, Speech and Signal Processing (ICASSP)*, 2014
12. M. Hamid, N. Bjorsell, S. Ben Slimane, Energy and eigenvalue-based combined fully-blind self-adapted spectrum sensing algorithm. IEEE Trans. Veh. Technol. **65**(2), 630–642 (2015)

13. J. Wojtiuk, Randomized Sampling for Radio Design. University of South Australia, School of Electrical and Information Engineering, Ph.D. dissertation, University of South Australia, School of Electrical and Information Engineering, 2000
14. I. Bilinskis, A. Mikelsons, *Randomized Signal Processing* (Prentice Hall, Cambridge, 1992)
15. Y. Zeng, Y.C. Liang, Spectrum-sensing algorithms for cognitive radio based on statistical covariances. IEEE Trans. Veh. Technol. **58**(4), 1804–1815 (2009)
16. F.J. Beutler, Error-free recovery of signals from irregularly spaced samples. SIAM Rev. **8**(3), 328–335 (1966)

Chapter 15
Prevention of Flooding Attacks in Mobile Ad Hoc Networks

Gurjinder Kaur, V. K. Jain, and Yogesh Chaba

15.1 Introduction

A mobile ad hoc network (MANET) is a wireless local area network model without the need of central base stations and operates as a self-organized, dynamically changing multi-hop network [1]. Applications of MANET's are most beneficial in medical emergencies, during natural catastrophes, for military applications, and conducting geographic exploration [2]. Devices used in a MANET are usually called mobile nodes. These nodes are characterized by high mobility, low power, limited storage, limited transmission range, and finite energy [3]. Mobile nodes communicate through bi-directional radio links within limited transmission range [4]. MANET in spite of having numerous advantages suffers from few disadvantages. Data transmission and security are the key challenges in MANET communication and need to be addressed in order to optimize the performance of MANET [5].

Routing in MANET is basically based on a hop-by-hop routing scheme. In MANET communication between source and destination is carried out through intermediate nodes and may involve traversing through several hops. Ad hoc on-demand vector (AODV) routing protocol is most preferred in MANET because it offers quick adaption to dynamic links, low processing, and memory overhead [6]. When a source node wants to communicate with destination a valid route to destination is required. A route request (RREQ) message is disseminated to its neighbours. Each node after receiving the message creates a reverse route to the source. RREQ message is flooded until the required route to destination is found

---

G. Kaur (✉) · V. K. Jain
Sant Longowal Institute of Engineering and Technology, Longowal, India

Y. Chaba
Guru Jambheshwar University of Science and Technology, Hisar, India

© Springer Nature Switzerland AG 2019
I. Woungang, S. K. Dhurandher (eds.), *2nd International Conference on Wireless Intelligent and Distributed Environment for Communication*, Lecture Notes on Data Engineering and Communications Technologies 27,
https://doi.org/10.1007/978-3-030-11437-4_15

by either reaching the destination or reaching an intermediate node that has a valid route (in its routing table) to the destination. A route reply (RREP) message is sent back to the source in a unicast manner through the reverse route created by nodes to source. Duplicate copies of RREQ packets received at any node are discarded. Once the RREP is received by source node, transmission of data packets is started to destination [7].

Most of the security breach attacks are initiated in MANET during route discovery process. Any legal node after receiving RREQ first time in an AODV-based MANET has the compulsion to re-disseminate the message if it is not destination and does not have valid route to destination. AODV is designed for use in networks where the nodes can all trust each other, either by use of preconfigured keys or because it is known that there are no malicious intruder nodes [8]. This trust based communication aspect of AODV is often explored by intruders in order to initiate attacks in network. Most common attacks deployed during route discovery in MANET are blackhole, wormhole, grayhole, and flooding that leave a considerable negative impact on the performance of MANET. Routing table overflow and routing table poisoning attack are another type of routing attacks that severely effect on routing information stored in routing table. Out of the various attacks discussed as above, flooding attack is found to be most destructive because severity of flooding attack often leads to exhaustion of resources and bandwidth of network [9]. RREQ flooding, data flooding, and hello flooding are some of prominent flooding attacks that have been addressed in recent literature [10].

In this paper in order to confront the RREQ flooding attack, a new two-step protection method (TSPM) is proposed. In the first step protection is ensured from the outsider malicious nodes by validating the neighbours that attempt to intrude in network. While in the second step detection of insider malicious nodes is taken care off by monitoring the frequency of RREQs and other parameters mentioned in RFC 3561 [8]. The proposed TSPM is tested by performing extensive simulation through Qualnet 5.02 network simulator. The obtained simulation results demonstrate that the proposed defence scheme is successful in reducing the false RREQ traffic loads and subsequently leads to better performance of MANET.

The rest of this paper is organized as follows: Sect. 15.2 describes the background profile and related work regarding flooding attacks in MANET. In Sect. 15.3, we present and describe relevant constituent elements of the proposed TSPM implemented by spreading defensive loads among legal nodes in a co-operative manner. The network experiments performed with QualNet 5.02 network simulator are presented and analysed in Sect. 15.4. Section 15.5 presents the concluding remarks.

15.2 Related Work

In MANETs, nodes are receptive to being compromised easily because each node acts as an independent entity. Since it is difficult to hunt down the compromised nodes, attacks are far more damaging and harder to recover. Overall performance of MANET is deteriorated in terms of energy and data. One severe type of attack, namely, RREQ flooding attack in MANETs, was presented by [3]. Attack is initiated firstly by selecting fake IP addresses which are not inside the legal IP domain defined/configured in the target MANET. RREQ packets are issued containing such fake IP addresses into network. Obviously, there will be no response back to such RREQ packets. Significant time and energy are wasted and resources are bound to carry such false traffic in the absence of a strong security measures.

To resist RREQ flooding attack in MANETs [3] presented the flooding attack prevention (FAP) scheme. The neighbour suppression mechanism is proposed to handle the RREQ flooding attack and the path cut-off mechanism for the data flooding attack. In the neighbour suppression mechanism, priority of neighbouring nodes is determined by inversely proportion to its frequency of originating RREQ packets. The threshold is determined by the maximum number of originating RREQ packets in a certain time period. If threshold is violated by neighbouring node to forward RREQ packets, the receiving node simply denies them.

Threshold prevention is introduced to improve above discussed FAP by defining a fixed RREQ threshold. Node is considered malicious if the rate of RREQ packet exceeds the threshold value. The scheme is not effective if malicious node violates parameters other than threshold, i.e. TTL value, etc. A distributed approach to resist the flooding attack is introduced in [11]. Two threshold values, RATE LIMIT and BLACKLIST LIMIT, are defined in algorithm. A RREQ from a neighbour is processed only if the number of previously received RREQ from this neighbour is less than RATE LIMIT. On the other hand, if the number of previously received RREQ from this neighbour is greater than BLACKLIST LIMIT, the RREQ is discarded and this neighbour is blacklisted. If the number of previously received RREQ from this neighbour is greater than RREQ LIMIT and less than BLACKLIST LIMIT, the RREQ is queued for processing after a delay expires. A disadvantage of this approach is the ability of the attacker to subvert the algorithm by disseminating thresholds levels and the possibility of permanently suspending a blacklisted neighbour that is not malicious. Meanwhile, the avoiding mistaken transmission table (AMTT) Scheme [12] suggests a defence system against the malicious flooding attack by utilizing an avoiding mistaken transmission table. The AMTT scheme requires huge memory space and considerable processing time for saving the packets at each node. The detect and isolate malicious host (DIMH) [13] uses the topology information and the public key cryptosystem to detect colluding malicious nodes. However, it is very hard to utilize the key management and exchange in mobile ad hoc networks. Many other rate limit based [14–16], behaviour based, and trust based techniques [17] are discussed in literature.

Although the issue of tackling the flooding has been addressed by numerous researchers [18–20] but still there exists a need of a methodology which can take care of attackers who are part of MANET as well as foreign entities (attackers not part of MANET in consideration).

15.3 Two-Step Protection Method

In AODV, when data communication is to be initiated between source and destination but no route information is there with source, a route discovery process is initiated by flooding a RREQ packet. Each RREQ packet has a unique identifier so that nodes can detect and drop duplicate packets. An intermediate node, upon receiving RREQ, verifies the information stored in RREQ packet and if required, it records the previous hop and the source node information in its route table to create a backward route. It sends back a ROUTE REPLY (RREP) packet to the source if it has a route to the destination; otherwise, it broadcasts the RREQ packet to continue search for destination. Any node can generate the RREP, if it is itself the destination or it has route to destination. The destination node sends a RREP through the backward route stored during transmission.

Below are some points that should be taken care off because intruders try to exploit these parameters to breach the security and initiate attack in network.

- RREQ_RATELIMIT: A node should not originate more than RREQ_RATELIMIT RREQ packets per second, it is defined 10 according to RFC 3561 [8]. The first time a source node broadcasts a RREQ packet, it waits for NET_TRAVERSAL_TIME milliseconds. IF a RREP is not received within that time, the source node then is allowed to send a new RREQ packet up to maximum allowed RREQ_RETRIES (default value is 2).
- Binary Exponential Backoff: A binary exponential backoff is utilized to reduce congestion in network for repeated attempts by source node to find a route to destination. It is used to define the waiting time for RREP for next new trail of RREQ packet.
- Time to Live (TTL): The value of TTL is incremented with next new trail of RREQ. In an expanding ring search, the originating node initially uses a TTL = TTL_START mentioned in the RREQ packet IP header and sets the timeout for receiving a RREP to RING_TRAVERSAL_TIME milliseconds. If the RREQ times out without reception of RREP, the originator broadcasts the RREQ again with the TTL incremented by TTL_INCREMENT. This continues until the TTL set in the RREQ reaches TTL_THRESHOLD, beyond which a TTL = NET_DIAMETER is used for each attempt.
- Fake Destination IP address: Malicious nodes originate RREQ packets using out-of-domain destination IP address to elongate the search process in network. Resources of network are wasted for searching a node that actually does not exist. It is hard to detect out-of-domain destination IP address because MANETs are of ad hoc nature, any node is free to leave and join network at any time.

However malicious nodes often violate above mentioned parameter values. In case RREQ_RATELIMIT is overlooked, massive RREQ packets may be originated in high frequency using out-of-domain destination IP addresses. They would also resend RREQ packets without waiting for binary exponential backoff time, and even they may send excessive RREQ packets with maximum TTL value in first attempt.

Fig. 15.1 N_list format

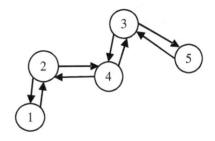

Fig. 15.2 Exchange of hello packets in MANET nodes

In extreme situation, the whole network is possibly congested with false RREQ packets flooded by malicious nodes.

In the first step of TSPM, each node maintains a list of legal neighbours (N_List) by periodically exchanging hello packets. IP address and status of neighbour node are recorded in N_List. Upon receiving any RREQ packet, each node validates the source node by checking IP address and status of source node in its N_List. If IP address is not found in the N_List, received RREQ is simply discarded and an attempt of illegal node to enter in network is protected.

In the second step violations of RREQ_RATELIMIT, binary exponential backoff, and TTL value are checked to hunt down a malicious node. Explanation of proposed technique is as follows.

A N_List as shown in Fig. 15.1 is designed to monitor each node originating RREQ packets, whether it is a legal or a malicious node. The structure of N_List is described below.

To get link availability of neighbour nodes of each legal node, TSPM utilizes the hello messages defined in [8]. Link availability information is updated by broadcasting local hello messages. A node uses hello message after every HELLO_INTERVAL milliseconds, whenever a node receives a hello message from a neighbour, the node should make sure that it has an active route to the neighbour and create one if necessary.

The values of N_IP address and link availability are recorded after receiving hello packet from neighbour. In field link availability value 1 indicates that the route is active and 0 indicates lost. The third field, node status, contains three values, derived from the second step of proposed technique. Values 1, 2, and 3 represent normal, suspicious, and malicious, respectively.

In Fig. 15.2, for example, there are two neighbour nodes 2 and 3 in transmission range of node 4. After exchanging the hello packets, IP addresses and link availability of nodes 2 and 3 will be recorded in N_List of node 4. Value 1 in link availability field for nodes 2 and 3 indicates that node 4 has two valid neighbour nodes.

Second step of proposed techniques initiates with the request of data transmission by any node. In case of unavailability of route to destination source node disseminates RREQ packet. Node status is updated after analysing RREQ received from any node. To check the legality of source node, source IP address of RREQ packet is compared with the addresses stored in N_List. If the same IP address does not exist in N_List, source node is considered malicious and node status field is updated with value 3 indicating malicious. Otherwise RREQ has to pass through the another check described in the second step of the proposed technique.

An intelligent attacker sometimes pretends to be a legal node by obeying the RREQ_RATELIMIT but cleverly put the maximum TTL value in the first attempt of RREQ with out-of-domain destination IP address. So it becomes challenging to detect the attacker by checking only RREQ_RATELIMIT. In the proposed technique TTL value and binary exponential backoff time are also checked additionally to chase the attacker. For RREQ_RATELIMIT two thresholds, soft and hard threshold values, are defined as per [3]. If rate of firing RREQ is above soft threshold but less than hard threshold value, node status for that node is updated to value 2; otherwise, if it is greater than hard threshold value node is considered as malicious and node status will be updated to 3. In case, rate of RREQ is within limit than other two parameters, i.e. TTL and binary exponential backoff, will be checked and accordingly node status field in Fig. 15.2 will be updated. After detecting the malicious node, a message including IP address of malicious node is broadcasted to inform all nodes in network.

15.4 Simulation and Result

TSPM is subjected to extensive simulation through Qualnet 5.02. The various parameters setting is presented in Table 15.1. Most effective metrics to measure the performance of routing in MANET are throughput, delay, routing overhead, and packet delivery ratio. Some other parameters like network overhead, network load, and energy consumed by network are also taken into consideration to measure performance of network. To analyse the proposed work the following metrics are used:

Table 15.1 Simulation parameters

Parameter	Value
Network simulator	QualNet 5.02
Simulation time	100 s
Items to send	100
No. of mobile nodes	50
No. of flooding nodes	1–5
Area	$1000 \times 1000 \text{ m}^2$
Routing protocol	AODV
Traffic	CBR
Packet size	512
Mobility model	Random way point
MAC protocol	IEEE 802.11

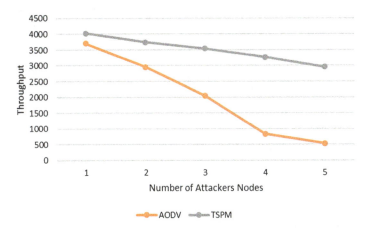

Fig. 15.3 Throughput vs number of attackers

- *Throughput:* It is defined as total amount of data in terms of number of bytes received by the destination per second.
- *Average End to End Delay:* The packet end to end delay is the average time of the packet passing through inside the network. It includes all over the delay of network like transmission time delay which occurs due to routing broadcasting and buffer queues. It also includes the time from generating packet from sender to destination and expressed in seconds.
- *Packet Delivery Ratio (PDR):* This is the ratio of the number of packets received at the destinations to the number of packets sent from the sources.

Different scenarios have been created using simulation parameters given in Table 15.1. The effectiveness of proposed technique in terms of throughput, PDR, and average end to end delay is analysed and is presented below.

Throughput Throughput is analysed with respect to increase in number of attackers with fixed number of nodes. It is clear from Fig. 15.3 that throughput decreases with increase in number of attacking nodes because flooding of RREQ packets drastically increases up to the capacity of bandwidth. Whole network clogs with the fake traffic and transmission of legal traffic is almost blocked. With the proposed TSPM a considerable improvement in throughput can be seen in Fig. 15.3.

Average End to End Delay As it can be seen in Fig. 15.4 that with increase in number of attacking nodes, average end to end delay also increases. By capturing all bandwidth with flooding of fake RREQ packets, attacking nodes forces the delay in genuine traffic. By using TSPM average end to end delay decreases because fake traffic is identified only within one hop, i.e. by the neighbouring node and network functions normally.

Packet Delivery Ratio As shown in Fig. 15.5, PDR decreases with increase in number of attackers. The effect of attack is less when one or two attackers exist in

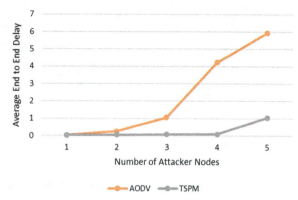

Fig. 15.4 Average end to end delivery vs number of attackers

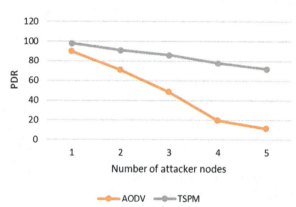

Fig. 15.5 Packet delivery ratio vs number of attackers

the network but with more than two attackers PDR drastically decreases. Flooding of RREQ by more attackers consumes the maximum bandwidth and other network resources. The proposed TSPM improves PDR significantly as compared to AODV protocol.

15.5 Conclusion

Mobile ad hoc networks exhibit vulnerabilities to malicious attacks because of their inherent characteristics. In this paper in order to confront the RREQ flooding attack in MANET, a new two-step protection method (TSPM) is proposed. The proposed TSPM is tested by performing extensive simulation through Qualnet 5.02 network simulator. The obtained simulation results demonstrate that the proposed defence scheme, namely, TSPM, is successful in reducing the false RREQ traffic loads and subsequently leads to better performance of MANET.

References

1. Y.C. Tseng, J.R. Jiang, J.H. Lee, Secure bootstrapping and routing in an IPv6-based ad hoc Network. J. Internet Technol **5**(2), 123–130 (2004)
2. H. Yang, H. Lou, F. Ye, S. Lu, L. Zhang, Security in mobile ad hoc networks: challenges and solutions. IEEE Wirel. Commun. **11**(1), 3847 (2004)
3. Y. Ping, D. Zhoulin, Z. Yiping, Z. Shiyong, Resisting flooding attacks in ad hoc networks, in *International Conference on Information Technology: Coding and Computing* (2005), pp. 657–662
4. F.-C. Jiang, C.-H. Lin, H.-W. Wu, Lifetime elongation of ad hoc networks under flooding attack using power-saving technique. Ad Hoc Netw. **21**, 8496 (2014)
5. E. Zamani, M. Soltanaghaei, The improved overhearing backup AODV protocol in MANET. J. Comput. Netw. Commun. **5**, 8 (2016)
6. C.E. Perkins, E.M. Royer, Ad-hoc on-demand distance vector routing. Presented at the Proceedings of the Second IEEE Workshop on Mobile Computer Systems and Applications, 1999
7. M. Elakkiya, N.E. Elizabeth, Opportunistic routing to forgo flooding attacks in MANET. Int. J. Eng. Dev. Res. 34–40 (2014)
8. C. Perkins, E. Belding-Royer, S. Das, *Ad hoc On-Demand Distance Vector (AODV) Routing. RFC 3561* (2003)
9. B. Kannhavong, H. Nakayama, Y. Nemoto, N. Kato, A. Jamalipour, A survey of routing attacks in mobile ad hoc networks. Wirel. Commun. **14**(5), 85–91 (2007)
10. R. Mishra, S. Sharma, R. Agrawal, Vulnerabilities and security for ad-hoc networks, in *2010 International Conference on Networking and Information Technology* (IEEE, Piscataway, 2010), pp. 192–196
11. J.-H. Song, F. Hong, Y. Zhang, Effective filtering scheme against RREQ flooding attack in mobile ad hoc networks. Presented at the Seventh International Conference on Parallel and Distributed Computing, Applications and Technologies, 2006
12. S. Li, Q. Liu, H. Chen, M. Tan, A new method to resist flooding attacks in ad hoc networks, in *2006 International Conference on Wireless Communications, Networking and Mobile Computing* (IEEE, Piscataway, 2006), pp. 1–4
13. Z. Xia, J. Wang, DIMH: a novel model to detect and isolate malicious hosts for mobile ad hoc network. Comput. Stand. Interfaces **28**(6), 660669 (2006)
14. P. Singh, A. Raj, D. Chatterjee, Flood tolerant AODV protocol. Int. J. Comput. Appl. **53**(6), 18–27 (2012)
15. Y. Ping, H. Yafei, Z. Yiping, Z. Shiyong, D. Zhoulin, Flooding attack and defence in ad hoc networks. J. Syst. Eng. Electron. **17**(2), 410–416 (2006)
16. K. Bhuvaneshwari, F.S. Devaraj, PDS- A profile based detection scheme for flooding attack in AODV based MANET. Int. J. Sec. Priv. Trust Manag. **2**(3), 17–28 (2013)
17. R. Venkataraman, M. Pushpalatha, T.R. Rao, Performance analysis of flooding attack prevention algorithm in MANETs. Int. J. Comput. Inf. Eng. **3**(8), 2056–2059 (2009)
18. M.J. Faghihniya, S.M. Hosseini, M. Tahmasebi, Security upgrade against RREQ flooding attack by using balance index on vehicular ad hoc network. Wirel. Netw. **23**(6), 1863–1874 (2017)
19. M.R. Hasan, Y. Zhao, Y. Luo, G. Wang, R.M. Winter, An effective AODV-based flooding detection and prevention for smart meter network. Procedia Comput. Sci. **129**, 454–460 (2018)
20. H. Kim, R.B. Chitti, J. Song, Handling malicious flooding attacks through enhancement of packet processing technique in mobile Ad Hoc networks. J. Inf. Process. Syst. **7**(1), 137–150 (2011)

Chapter 16
Exploiting ST-Based Representation for High Sampling Rate Dynamic Signals

Andrea Toma, Tassadaq Nawaz, Lucio Marcenaro, Carlo Regazzoni, and Yue Gao

16.1 Introduction

ST is a time–frequency technique which provides time information of the spectral content of the signal by observing it through a sliding Gaussian window. A major drawback of ST is heavy computational cost [11] especially in wideband applications.

This chapter introduces a dual-resolution (DR) approach based on discrete ST which reduces the computational time of the conventional ST by increasing time delay of the sliding window. Specifically, the vector signal is divided into sub-blocks in which the wideband signal is assumed to be locally stationary and the sliding window is moved on the first sample of each sub-block. This results in a control

---

A. Toma (✉)
Department of Electrical, Electronic, Telecommunications Engineering and Naval Architecture (DITEN), University of Genoa, Genova, Italy

School of Electronic Engineering and Computer Science (EECS), Queen Mary University of London, London, UK
e-mail: andrea.toma@ginevra.dibe.unige.it; a.toma@qmul.ac.uk

T. Nawaz · L. Marcenaro · C. Regazzoni
Department of Electrical, Electronic, Telecommunications Engineering and Naval Architecture (DITEN), University of Genoa, Genova, Italy
e-mail: tassadaq.nawaz@ginevra.dibe.unige.it; lucio.marcenaro@unige.it; carlo.regazzoni@unige.it

Y. Gao
School of Electronic Engineering and Computer Science (EECS), Queen Mary University of London, London, UK
e-mail: yue.gao@qmul.ac.uk

© Springer Nature Switzerland AG 2019
I. Woungang, S. K. Dhurandher (eds.), *2nd International Conference on Wireless Intelligent and Distributed Environment for Communication*, Lecture Notes on Data Engineering and Communications Technologies 27,
https://doi.org/10.1007/978-3-030-11437-4_16

capability on both time-resolution and computational load that allows a dynamic trade-off to be adaptively selected.

Applications of wideband dynamic spectrum sensing include TV white spaces (TVWS) [12], to address the shortage of the wireless spectrum and to utilize it efficiently, and PHY-layer security because radio communications in wireless environments experience external attacks from malicious devices owing to the broadcast nature of radio propagation [18]. In both cases, time–frequency analysis can provide information to analyse dynamically adapted transmitter spectrum occupancy strategies.

In this context, the DR technique is conceived to deal with dynamic wideband signals sampled at high sampling rate and to produce a better trade-off between time-resolution and computational time. The proposed approach has been validated in a dynamic scenario consisting of real data collected with a software-defined radio testbed. As described in Sect. 16.6, the main target for this work is signal identification in Cognitive Radio (CR) through features extracted from each detected signal in the time–frequency representation such as bandwidth, carrier frequency, amplitude, and variance. To this end, a comparison between the ST representation and the corresponding STFT of the analysed spectrum supports the choice of using ST.

The remainder of the chapter is organized as follows: Section 16.2 introduces the time–frequency framework; discrete ST and the DR approach are described in Sect. 16.3; the testbed and data acquisition are in Sect. 16.4; while the validation and comparison with STFT are detailed in Sect. 16.5; and some future directions conclude the chapter.

16.2 Time–Frequency Transforms

Since the well-known Fourier transform (FT) assumes that the signal is stationary, i.e., its frequency content is constant over the time, it is unsuitable for analysing non-stationary waveforms. Consequently, time–frequency (TF) analysis is found in many applications such as speech processing, astronomy, or medicine [13] which allows signals characterized by non-stationary behaviours to be analysed and locally consistent temporal information about their spectral content to be represented.

Wigner-Ville distribution (WVD) is defined as Fourier transform of a local correlation function called bilinear transform [3]. Since time resolution and frequency resolution of WVD are independent of each other, WVD shows high aggregation properties. On the other hand, it results in an unavoidable cross-term factor. In Hilbert–Huang transform (HHT) method [9], a multi-component non-stationary signal is decomposed by using the empirical mode decomposition (EMD) technique and, then, the Hilbert transform (HT) is applied to the obtained components known as intrinsic mode functions (IMFs). Although the decomposition is adaptive and therefore highly efficient, IMFs suffer from mode mixing issue that results in an improper time–frequency representation.

16 Exploiting ST-Based Representation for High Sampling Rate Dynamic Signals

The ST, developed by Stockwell et al. [15] and based on a sliding Gaussian window distribution, exhibits globally referenced phase and frequency measurements similar to those of the STFT [1], as well as the progressive resolution of the wavelet transform (WT) [8]. Despite this combination of desirable features, the computational demands of the ST in some cases limit its utility and prevent more widespread usage. One of the most recent applications extensively employing the ST is Automatic Modulation Recognition (AMR) in Cognitive Radio [16].

In this chapter, a DR approach applied to ST is proposed, and the corresponding representations are compared with the ones obtained through STFT. The latter is computed by applying the discrete Fourier transform (DFT) through a fixed-sized, moving window to a given time series $\{r_p\}_{p=0}^{P-1}$ of finite length P [10]. The window is moved by one time point at time resulting in overlapping windows. For a rectangular window of length N, the STFT representation consists of N frequency points and $P - N + 1$ time points. The implementation of the STFT matrix can be found in [10]. Compared with the ST, STFT exhibits less computational complexity. However, as shown in Sect. 16.5, the corresponding representations are not sufficient for possible application to signal identification through features extracted from detected waveforms.

16.3 Discrete ST with Dual-Resolution

The main objective of this section is to introduce the proposed approach for the discrete model of the ST distribution. More specifically, $r[p]$, $p = 0, 1, \ldots, P-1$, denotes the discrete time series corresponding to a continuous signal $r(t)$ with a time sampling interval T. The discrete ST of $r[p]$ is given by [2]:

$$S_T[m, n] = \sum_{p=0}^{P-1} r[p] \frac{|n|}{\sqrt{2\pi kN}} e^{-\left(n^2(m-p)^2/2k^2N^2 + j2\pi pn/N\right)} \tag{16.1}$$

when $m = 0, \ldots, M - 1$; $n = 1, \ldots, N - 1$; and by:

$$S_T[m, 0] = \frac{1}{P} \sum_{p=0}^{P-1} r[p] \tag{16.2}$$

when $n = 0$; m is the time delay of the sliding window, n denotes the index of frequency range, p denotes the time index, and k is a scaling factor that controls the time–frequency resolution. When k increases, the frequency resolution increases, with a corresponding loss of time resolution [13].

When $m = 0, \ldots, M - 1$ and $n = 0, \ldots, N - 1$ (with $M = N = P$ from now on), the number of time and frequency samples after the ST is exactly $P \cdot P$ and, consequently, the computational load is limited by the number of the samples,

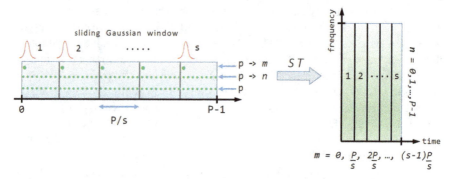

Fig. 16.1 Dual-resolution technique based on ST

P, in the signal. Considering that m is the time delay of the sliding window over the signal $r[p]$ and assuming that the signal is locally stationary within the time corresponding to several time delays m, a different approach is proposed in this chapter which reduces the computational load of the discrete ST. From Fig. 16.1, the P samples of the locally stationary signal are divided into s sub-blocks each consisting of P/s samples.

This approach computes the ST on s equally spaced values for m belonging to the set $\{0, \ldots, P-1\}$, while $n = 0, \ldots, P-1$ remains as before. Indeed, since in this way the time delay of the sliding window is increased, all the values of m in each sub-block (unless the first one) are not directly included in the computation of the ST (it is sufficient to increase the sliding window length, by increasing k, to also cover the discarded samples). This is based on the consideration that the signal is locally stationary over P/s m's, or (P/s)-stationary, namely variations of the parameters happen no faster than the time corresponding to P/s samples and not necessarily at the boundary between two consecutive sub-windows. For quicker changes, the length of the sub-blocks can be reduced by increasing s accordingly. This introduces a new capability of the ST, namely a trade-off between time-resolution and computational time that can be adaptively controlled through s, P, and k. The frequency resolution can be changed by regulating P or k (or both).

Equations (16.1) and (16.2) can be written in a vectorial form as follows:

$$S_T[m,n] = \mathbf{T}_{mn}\mathbf{r} \tag{16.3}$$

where the $P \times 1$ vector $\mathbf{r} = [\,r[0], r[1], \ldots, r[P-1]\,]^T$ is the discrete series, while the following $1 \times P$ vector:

$$\mathbf{T}_{mn} = [\,T_{mn,0}, T_{mn,1}, \ldots, T_{mn,(P-1)}\,]$$

consists of elements given by [2]:

$$T_{mn,p} = \frac{|n|}{\sqrt{2\pi}kN} e^{-(n^2(m-p)^2/2k^2N^2 + j2\pi pn/N)} \tag{16.4}$$

16 Exploiting ST-Based Representation for High Sampling Rate Dynamic Signals

$p = 0, \ldots, P - 1$. By using the $(s \cdot N) \times P$ transform matrix

$$\mathbf{T} = \left[\; \mathbf{T}_{00}^T, \ldots, \mathbf{T}_{0(N-1)}^T, \mathbf{T}_{\frac{P}{s}0}^T, \cdots \right.$$

$$\left. \cdots, \mathbf{T}_{\frac{P}{s}(N-1)}^T, \cdots, \mathbf{T}_{\frac{P(s-1)}{s}0}^T, \cdots, \mathbf{T}_{\frac{P(s-1)}{s}(N-1)}^T \; \right]^T$$

the discrete ST can be modelled as a linear vectorial equation:

$$\mathbf{s} = \mathbf{Tr} \qquad (16.5)$$

in which the $(s \cdot N) \times 1$ vector $\mathbf{s}$ consists of elements s_{mn} corresponding to $S_T[m, n]$. In this chapter, the dimensionality of Eq. (16.5) is drastically reduced because $m = m_s P/s$ where $m_s = 0, \ldots, s - 1$ is the delay index, instead of $m = 0, 1, 2, \ldots, P - 1$.

Now, let B be the frequency bandwidth of the signal of interest, which in practical cases is from several MHz to GHz for wideband signals. Consequently, the sample rate is in the order of $2B$, or slightly more. (P/s)-stationarity of the signal means that is should be stationary in a time length of $\frac{P/s}{2B}$ s. In Sect. 16.5, P/s is from 4 to 64. Consequently, $\frac{P/s}{2B}$ is from about 16 to 267 ns for $B = 120\,\text{MHz}$. While, it is from 2 to 32 ns when $B = 1\,\text{GHz}$. Although the piecewise assumption is a limitation of the DR approach, the time interval in which the signal should be stationary is very short and compatible with real systems.

Concerning the frequency resolution defined as $R = B/P$ which should be a small value, P must be as large as possible for wideband signals in order to have reasonable resolution. This motivates the DR approach.

16.4 Spectrum Measurements on Real Dynamic Signals

This section describes both the SDR testbed used to generate real dynamic signals and the data acquisition process.

16.4.1 Testbed Architecture

The testbed is an SDR platform, Fig. 16.2, which consists of two Secure Wideband Multi-role—Single-Channel Handheld Radios, SWAVE HHs, connected through a dual-directional coupler [6]. A SWAVE HH radio is capable of generating the digital SelfNET Soldier Broadband Waveform (SBW) with bandwidth up to 5 MHz [5] and provides operability in both Very High Frequency band, VHF (30–88 MHz), and Ultra High Frequency band, UHF (225–512 MHz).

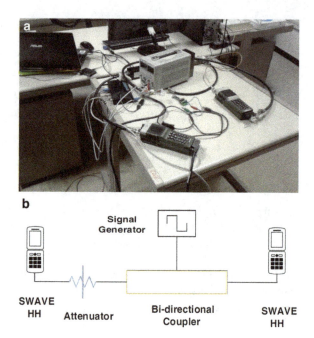

Fig. 16.2 SDR testbed utilized to generate dynamic signals in a wideband spectrum: (**a**) hardware platform, and (**b**) diagram of the main components of the testbed and their connections

The testbed provides support for remote control of HH's transmit and receive parameters through Ethernet and the Simple Network Management Protocol (SNMP v3) [6].

In the spectrum sensing process, the HH's 14-bit Analog-to-Digital-Converter (ADC) performs sampling at 250 Msamples/s. Every 3 s, a burst of 8192 consecutive complex samples is output over the RS-485 serial port. The bandwidth of the corresponding spectrum is 120 MHz, and the effective resolution is 29.3 kHz/sample. Full details on the testbed architecture can be found in [5].

16.4.2 Data Acquisition

By means of the testbed described in the previous section, real data is collected in the frequency band 0–120 MHz. As shown in Fig. 16.3 (thin dotted blue line), each received spectrum consists of a digitally modulated SBW signal in the VHF band (from the transmitting HH) and a number of signals (from the environment) such as the FM signal (in the 88–108 MHz band) and the baseband (BB) signal at 0–7 MHz, in addition to interference (at 20 and 80 MHz) and noise. The SBW signal is capable of hopping within the VHF band.

The intermediate step before ST analysis is preprocessing in the frequency domain which is applied to the data in order to: reduce the fluctuations in the spectrum by smoothing, remove the noise and detect the signals through energy

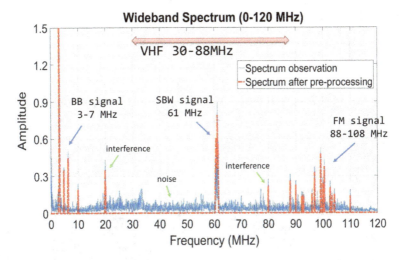

Fig. 16.3 Wideband spectrum, which consists of a number of signals including the SBW at 61 MHz, before preprocessing (thin dotted blue line) and after preprocessing (thick dash-dotted red line)

detector, group the FFT bins which belong to the same signal, and smooth the grouped bins [7]. The resulting spectrum observation is shown in Fig. 16.3 (thick dash-dotted red line).

16.5 Validation and Comparison with STFT

To validate the proposed dual-resolution approach, the discrete ST has been applied to real data described is Sect. 16.4.2. Specifically, at four different time instants the carrier frequency of the transmitted SBW signal assumes sequentially the values 41–51–61–71 [MHz], while the transmit power is 7 dBm. The other waveforms extracted from the spectrum shown in Fig. 16.3 include: 4 sub-signals close to one another (spacing about 0.5–1 MHz) at low frequencies, 2 interference signals at 20 MHz and 80 MHz, respectively, and 3 sub-signals in the FM band (spacing about 2.5 and 7 MHz). The carrier frequency of these waveforms is fixed over the measurement time.

According to Table 16.1, the analysed parameters are P (number of samples in the signal to be S-transformed), s (number of sub-blocks), and the scaling factor k described in Sect. 16.3. The SBW signal is at least (P/s)-stationary for each configuration.

Validation is performed through Matlab® 2106b. Figure 16.4a is the conventional ST, obtained with $k = 17.5$ and without DR, of a signal which consists of $P = 512$ samples. m_s is the delay index as defined in Sect. 16.3, namely the time

Table 16.1 Parameters used to validate the dual-resolution approach

P				s			k		
256	512	640	1024	16	32	64	5	17.5	25

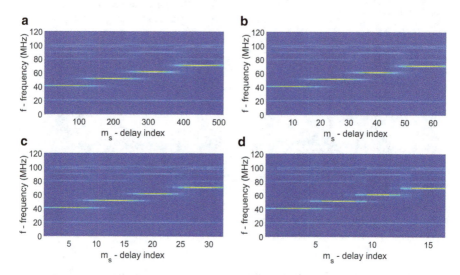

Fig. 16.4 ST representation of the wideband signal with $P = 512$ and $k = 17.5$: (**a**) without DR; and with dual-resolution where (**b**) $s = 64$, (**c**) $s = 32$, and (**d**) $s = 16$. The SBW jumps among 41–51–61–71 MHz at different time instants

variable after ST. The waveforms in the spectrum can be clearly distinguished. In particular, the SBW signal in the middle of the figure jumps sequentially to the four different frequencies at four consecutive time instants. At the bottom, there are the BB peaks (they are very close to each other; the progressive resolution of ST produces fine frequency resolution at low frequencies [4]). Just above them, there is the interference at 20 MHz. Beyond the SBW signal, the interference at 80 MHz and the 3 FM sub-signals can be seen (the progressive resolution of ST produces low-frequency resolution at high frequencies [4]). In this case, the amount of frequency-time samples is $P \cdot P$. The dual-resolution technique generates the ST representation shown in Figs. 16.4b–d in which s is 64, 32, and 16, respectively. Although the time resolution gets worse by reducing the number of sub-blocks, the signals in the spectrum remain clear. In particular, the dynamic hopping of the SWB signal can be still observed. The main advantage is that the amount of frequency-time samples is reduced to $P \cdot s$, with $s \ll P$.

As shown in Fig. 16.5, signals can be detected by thresholding on the ST representation to obtain occupied areas in both the frequency and time domains. For the sake of clarity, the 3D representation corresponding to Fig. 16.4c is shown in this figure. These occupied regions correspond to locations where the wideband signal, such as the one represented in Fig. 16.5a, is above a certain threshold which is set as k times the mean value of the ST representation computed on a 2-dimensional basis;

16 Exploiting ST-Based Representation for High Sampling Rate Dynamic Signals

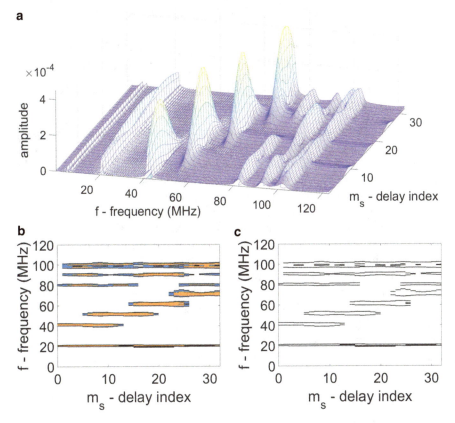

Fig. 16.5 Signal detection and contours extracted from ST representation of the wideband signal with $P = 512$, $k = 17.5$, and $s = 32$: (**a**) 3D representation, (**b**) 2D representation of energy locations detected by threshold $= 2\mu_{ST}$ (external dark-blue areas) and by threshold $= 4\mu_{ST}$ (internal light-orange areas), and (**c**) contours of the energy locations with threshold $= 2\mu_{ST}$. The SBW jumps among 41–51–61–71 MHz in the VHF band at different time instants

k is a positive factor used to tune the threshold. In the example below, $k = \{2, 4\}$ and the threshold is denoted as μ_{ST}. In other words, let **X** be the ST representation as in Fig. 16.5a. The threshold is computed as:

$$\text{threshold} = k \cdot \mu_{ST} = k \cdot \text{mean2}(\mathbf{X})$$

where mean2($\cdot$) is the mean function applied across two directions, namely time and frequency, and the outcome is a scalar value. Consequently, contours of the occupied regions for each signal, that are detected in the observed spectrum, can be extracted from the time–frequency representation of the wideband signal. An example is given as follows:

1. 3D representation with DR ($s = 32$) of the observed 0–120 MHz spectrum is shown in Fig. 16.5a;
2. Energy locations obtained with threshold = $2\mu_{ST}$ and with threshold = $4\mu_{ST}$ are shown in Fig. 16.5b (external dark-blue areas and internal light-orange areas, respectively); μ_{ST} is the mean value of the ST representation with DR;
3. Contours of the occupied regions detected with threshold = $2\mu_{ST}$ are shown in Fig. 16.5d.

In the middle of each figure, the non-stationary SBW signal can be easily seen which jumps in the VHF frequency range at frequencies 41–51–61–71 MHz where the signal is found at different time instants. This kind of analysis lays the foundations for extraction of dynamic features [17] from the 2-dimensional contours of each moving signal such as bandwidth, central frequency, transmitting power, and variance, as explained in the following section.

A comparison with the STFT shows the grade of applicability to a high sampling rate framework and, in particular, to possible signal identification through features extracted from each of the signals detected in the time–frequency representation. Specifically, the obtained STFT representation is shown in Fig. 16.6a–d for four different window lengths ($N = 128, 64, 32, 16$). A sensible choice for the length of the window, N, could reduce both the time resolution drop, happening when N is large, and the frequency resolution drop, happening when N is small. In any case, the STFT representations of wideband non-stationary signals are not sufficient for signal identification applications where high-accuracy discriminative features are requested.

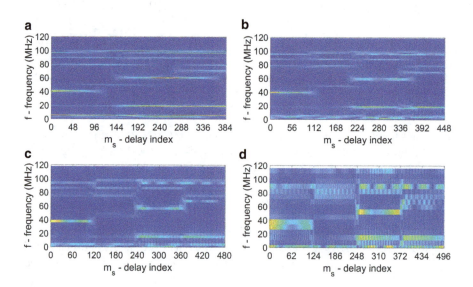

Fig. 16.6 STFT representation of the wideband signal with $P = 512$. The length of the sliding window is: (**a**) $N = 128$, (**b**) $N = 64$, (**c**) $N = 32$, and (**d**) $N = 16$. The SBW jumps among 41–51–61–71 MHz at different time instants

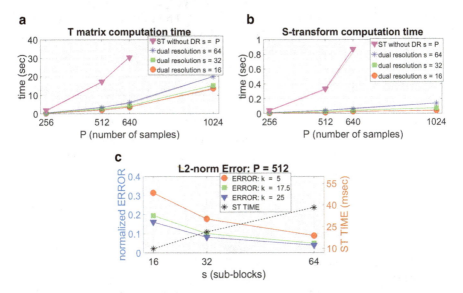

Fig. 16.7 Performance indicators: (**a**) time to generate **T**, (**b**) ST computation time, and (**c**) normalized $L2$-norm error

To show the effective benefits obtained with the proposed algorithm, Fig. 16.7 illustrates three different indicators: the time to create the **T** matrix, the time to compute the ST through Eq. (16.5), and the $L2$-norm error which illustrates the difference between the 2D TF representation without DR and the corresponding representation with DR ($s = 64, 32, 16$). Specifically, let ϵ_n be the normalized $L2$-norm error defined as $\frac{\|\mathbf{X}_s - \mathbf{X}_P\|_2}{\|\mathbf{X}_P\|_2}$, where the matrix $\mathbf{X}_s$ is the expanded ST representation with DR, obtained through P/s replicas of each column, the matrix $\mathbf{X}_P$ is the ST representation without DR, and $\|\cdot\|_2$ is the $L2$-norm function.

Figure 16.7a, b shows that in the conventional ST without DR the computational time to both generate the **T** matrix and perform the ST increases sharply when P is increased from 256 to 640. With $P = 1024$, Matlab is no longer capable of generating **T** because of its large dimensions ($P \cdot P$ rows and P columns). While, the dual-resolution approach reduces dramatically the computational time and $P = 1024$ is also feasible. This means that the frequency resolution can be improved. Basically, k doesn't influence the computational time which is though increased by s.

Figure 16.7c, where $P = 512$, shows that both s and k influence ϵ_n. As expected, the error (mostly related to the time resolution) increases when at least one between s and k decreases and it can be reduced by increasing k when s is small. In Fig. 16.7c, the corresponding ST computation time is also included which is inversely proportional to the error. The same results hold when a different value for P is considered.

To know the percentage of improvement, both the computation time and the time to number of samples ratio for the proposed approach are compared with the corresponding values for the conventional ST. From Fig. 16.7a, the 'T matrix computation time' is reduced from 17.1 s (ST without DR) to 1.9 (or 3.4) s (with DR), when $P = 512$. This corresponds to

$$\frac{\Delta \sec}{17.1} 100 \ (\%) = \frac{17.1 - 1.9 \ (\text{or } 3.4)}{17.1} 100 \ (\%) = \frac{15.2 \ (\text{or } 13.7)}{17.1} 100 \ (\%) = 88.89 \ (\text{or } 80.12)\%.$$

In other words, when $P = 512$, just $\frac{1.9 \ (\text{or } 3.4)}{17.1} 100 \ (\%) = 11.11 \ (\text{or } 19.88)\%$ of the time necessary to generate the **T** matrix without DR is requested when DR is applied. In addition, 30.2 s (ST without DR) and 3.5 (or 5.8) s (with DR) have been measured when $P = 640$. This corresponds to

$$\frac{\Delta \sec}{30.2} 100 \ (\%) = \frac{30.2 - 3.5 \ (\text{or } 5.8)}{30.2} 100 \ (\%) = \frac{26.7 \ (\text{or } 24.4)}{30.2} 100 \ (\%) = 88.41 \ (\text{or } 80.79)\%.$$

Namely, when DR is applied, just $\frac{3.5 \ (\text{or } 5.8)}{30.2} 100 \ (\%) = 11.59 \ (\text{or } 19.21)\%$ of the time necessary to generate the **T** matrix without DR is requested.

In the same picture, the time to number of samples ratio is $\frac{15}{1024} = 14.65 \times 10^{-3}$ for $P = 1024$ and $s = 32$ (with DR) and $\frac{30}{640} = 46.88 \times 10^{-3}$ for $P = 640$ and without DR; namely, more than three times smaller in ST with DR with respect to the conventional ST. Effectively, when DR is applied, the amount of samples to be ST-transformed can be doubled while guaranteeing comparable computational time to obtain the **T** matrix with conventional ST.

Similar conclusion results when values in Fig. 16.7b for 'ST computation time' are considered. Specifically, 0.33 s (ST without DR) in comparison with 0.01 (or 0.04) s (with DR), when $P = 512$ produces 3.03 (or 12.12)% which is the amount of necessary time to compute the ST with DR with respect the conventional ST. When $P = 640$, the measured time is 0.87 s (ST without DR) and 0.025 (or 0.06) s (with DR), corresponding to 2.87% and 6.90%, respectively.

In this case, the time to number of samples ratio is $\frac{0.07}{1024} = 68.36 \times 10^{-6}$ for $P = 1024$ and $s = 32$ (with DR) and $\frac{0.87}{640} = 1.36 \times 10^{-3}$ for $P = 640$ and without DR; namely, about 20 times smaller in ST with DR with respect to the conventional ST. Effectively, when DR is applied, the number of samples that guarantees comparable ST computational time, with respect to the case without DR, is much larger; till four times larger, when $s = 32$.

16.6 Conclusion and Future Directions

In many applications, such as dynamic spectrum sensing and dynamic spectrum access, processing wideband signals sampled at high sampling rate is the main objective. Both time and frequency information should be contemporary processed to represent the dynamic nature of the spectrum.

The novel ST-based dual-resolution approach proposed in this chapter overcomes the limitations of the conventional ST when applied to wideband signals by increasing the time delay of its sliding window and provides a trade-off between time-resolution and computational time which can now be regulated in real-time applications. Applied to real data, the ST with DR has shown a significant reduction of the computational time with respect to the case without DR, both to generate the matrix **T** and to calculate the ST, with just a slight decrease in time-resolution and TF representation accuracy. In particular, just 11–20% of the time necessary to generate the **T** matrix without dual-resolution is requested and 3–12% of the ST computational time with respect to the conventional ST. In addition, discrete series with more samples, which usually happens with wideband signals, can also be processed. Indeed, the number of samples can be till four times larger while guaranteeing comparable computational time.

A comparison between ST and STFT time–frequency representations shows that the STFT representations of wideband non-stationary signals are not sufficient to guarantee reliability in applications, like signal identification, where high-accuracy discriminative features extracted from each of the signals are requested. In this chapter, the concept of reliability is mainly related to features that can be extracted from TF representation; at this stage, a qualitative definition of this concept is considered. Consequently, from representations in Fig. 16.6, it results in a reduced quality of the images and, then, of the features that can be extracted, at least in one of the two domains (time/frequency). In future work, where extracted features are effectively employed, how much reliability is attained will be demonstrated by quantifying the concept through metrics related to the implemented algorithm such as accuracy in the classification rate when, for example, a classifier is applied to recognize the different signals inside the spectrum of interest. In signal identification, the rate of identification and its accuracy (based on the extracted features) is also a quantitative approach for the concept of reliability.

Moreover, the length of the sliding window in STFT should be chosen very accurately, mostly due to the time–frequency resolution trade-off. Since the proposed approach is meant to be applied in a context where information about dynamics of signals should be extracted from the TF representation, optimization of rate of detection of jumps or actions, rate of correct (re-)identification of signals when they appear somewhere in the spectrum, or accuracy of trajectories would be possible metrics that allow us to set a suitable length of the sliding window of the STFT. Results from the comparison made ST as the preferred choice for developing future work based on modelling the dynamics of signals.

Future work also assumes the presence of non-stationarities within sub-blocks and detects them to reduce the basic assumption of the proposed approach, and combines the DR with Discrete Orthogonal S-Transform (DOST) for a more efficient representation of ST [14]. In addition, dynamic models and learning mechanisms, based on ST computational gain and precision of detecting hop frequencies, can be implemented as part of Learning Dynamic Jamming models toward cognitive dynamic systems (CDSs) for PHY-layer security and Cognitive Radio.

Specifically, both time and frequency information are obtained from dynamic signals with controllable resolution in both the domains. The corresponding contours are then extracted. Future directions in this framework could include application of techniques and algorithms used for data analysis, image and video processing, robotics, and so forth, to wireless communications. Indeed, general concepts such as entity, state, and trajectory can be specified in CR as wireless signal, amplitude-central frequency-bandwidth-variance, and time-frequency information, respectively [17]. Considering that, dynamic models are learned from the observed features to represent moving signals through probabilistic distributions.

Validation of the ST-based DR will also be performed on TVWS signals in the 470–790 MHz band in the UK.

References

1. J.B. Allen, L.R. Rabiner, A unified approach to short-time Fourier analysis and synthesis. Proc. IEEE **65**(11), 1558–1564 (1977)
2. Y. Baiqiang, H. Yigang, A fast matrix inverse s-transform algorithm for MCG denoise, in IEEE International Conference on Electronic Measurement and Instruments (ICEMI), (2015), pp. 315–319
3. B. Bouachache, P. Flandrin, Wigner-Ville analysis of time-varying signals, in IEEE International Conference on Acoustics, Speech, and Signal Processing, ICASSP, Paris, France (1982), pp 1329–1332
4. R. Brown, M. Lauzon, R. Frayne, A general description of linear time-frequency transforms and formulation of a fast, invertible transform that samples the continuous s-transform spectrum nonredundantly. IEEE Trans. Signal Process. **58**(1), 281–290 (2010)
5. K. Dabčević, L. Marcenaro, C.S. Regazzoni, SPD-driven smart transmission layer based on a software defined radio test bed architecture, in 4th International Conference on Pervasive and Embedded Computing and Communication Systems (PECCS) (2014)
6. K. Dabčević, M. Mughal, L. Marcenaro, C.S. Regazzoni, Spectrum intelligence for interference mitigation for cognitive radio terminals, in Wireless Innovation Forum European Conference on Communications Technologies and Software Defined Radio (WInnComm- Europe) (2014)
7. K. Dabčević, M. Mughal, L. Marcenaro, C.S. Regazzoni, Cognitive radio as the facilitator for advanced communications electronic warfare solutions. J. Signal Process. Syst. **83**(1), 29–44 (2016)
8. I. Daubechies, The wavelet transform, time-frequency localization and signal analysis. IEEE Trans. Inf. Theory **36**(5), 961–1005 (1990)
9. N.E. Huang, et al., The empirical mode decomposition and Hilbert spectrum for nonlinear and nonstationary time series analysis. Proc. R. Soc. Lond. A Math. Phys. Eng. Sci. **454**(1971), 903–995 (1998)
10. S. Okamura, The short time Fourier transform and local signals, in Dissertations, 58. Department of Statistics, Carnegie Mellon University Pittsburgh, PA (2011)
11. K. Patel, N.C. Kurian, N.V. George, Time frequency analysis: a sparse S transform approach, in 2016 International Symposium on Intelligent Signal Processing and Communication Systems (ISPACS) (2016), pp. 1–4
12. Z. Qin, Y. Gao, M. Plumbley, C. Parini, Wideband spectrum sensing on real-time signals at sub-Nyquist sampling rates in single and cooperative multiple nodes. IEEE Trans. Signal Process. **64**(12), 3106–3117 (2016)

13. C. Simon, S. Ventosa, M. Schimmel, A. Heldring, J. Danobeitia, J. Gallart, A. Mànuel, The S-transform and its inverses: side effects of discretizing and filtering. IEEE Trans. Signal Process. **55**(10), 4928–4937 (2007)
14. R. Stockwell, A basis for efficient representation of the S-transform. Digital Signal Process. **17**(1), 371–393 (2007)
15. R. Stockwell, L. Mansinha, R. Lowe, Localization of the complex spectrum: the S transform. IEEE Trans. Signal Process. **44**(4), 998–1001 (1996)
16. M.V. Subbarao, P. Samundiswary, An intelligent cognitive radio receiver for future trend wireless applications. Int. J. Comput. Sci. Inf. Secur. **14**, 7–12 (2016)
17. A. Toma, C.S. Regazzoni, L. Marcenaro, Y. Gao, Learning dynamic jamming models in cognitive radios, in Handbook of Cognitive Radio. Cognitive Radio Applications and Practices (Springer, Berlin, 2018)
18. Y. Zou, X. Wang, L. Hanzo, A survey on wireless security: technical challenges, recent advances and future trends. Proc. IEEE **104**(9), 1727–1765 (2016)

Chapter 17
Real-Time Spectrum Occupancy Prediction

S. A. Abdelrahman, Omar Khaled, Amr Alaa, Mohamed Ali, Injy Mohy, and Ahmed H. ElDieb

17.1 Introduction

Cognitive radio (CR) is one of the empowering technologies of emerging wireless communication systems that addresses the problem of spectrum scarcity by making radios truly intelligent, allowing us to use the radio spectrum efficiently. CR has the capability of saving the wasted bandwidth, optimizing the underutilized spectrum, enhancing the overall throughput, and handling the scarcity of radio spectrum, the most tightly regulated resources of all time, through intelligent and cognitive capabilities and dynamic spectrum access and management policies. This novel approach and the term cognitive radio was first introduced by Joseph Mitola [1].

A CR should incessantly observe and learn about the surrounding radio frequency (RF) environment by continuously observing and extracting information from the RF domain, and get to know environment's and user's daily patterns and model the RF scene over space, and time, so it can respond and adapt to any dynamic change. Finally, a CR should be capable of predicting the upcoming changes in the RF domain and creating a high level of spectrum awareness so it can maximize its throughput and in the meantime reduce interference to target band of interest [2].

A major challenge in cognitive radio is that the secondary users need to detect the presence of primary users in the licensed spectrum and quit the frequency band

---

S. A. Abdelrahman (✉)
Scuola Superiore Sant'Anna, Pisa, Italy

O. Khaled · A. Alaa · M. Ali · I. Mohy · A. H. ElDieb
CIC - Canadian International College, Cairo, Egypt

© Springer Nature Switzerland AG 2019
I. Woungang, S. K. Dhurandher (eds.), *2nd International Conference on Wireless Intelligent and Distributed Environment for Communication*, Lecture Notes on Data Engineering and Communications Technologies 27,
https://doi.org/10.1007/978-3-030-11437-4_17

as quickly as possible if the corresponding primary radio emerges in order to avoid interference to primary users, which is achieved by deploying accurate and reliable spectrum sensing methods [3].

We have to think of cognitive radios in terms of software-defined radio (SDR) and cognitive engine (CE). SDR is a radio in which some or all of the physical layer parameters like gain, frequency, or waveform generation are software defined [4]. On the other hand, CE is the main logic which controls intelligent radio behavior.

Our aim in this chapter is to design a neural network-based energy detection for a higher level of prediction for over-the-air transmission on a larger scale that is outside of the traditional simulation environment. This is achieved because of the wide availability of open access testbeds which are dedicated to researchers and students for testing and developing real-time communication and wireless experiments. In our experiment, we used CORNET testbed [5] to develop and test our systems using Cognitive Radio Test System (CRTS) framework [6].

17.2 Spectrum Sensing Methods Overview

Spectrum sensing is the first cognitive cycle which plays the most important role in CR. The aim of spectrum sensing is to enable CR to detect spectrum holes and to coexist and share the spectrum with other users without harmful interference. In order to coexist, share the spectrum, and utilize the band effectively without interference, various detection methods have been discussed in [7–9], such as coherent detection schemes, like cyclostationary feature detection and matched filter, and non-coherent detection schemes, like energy detection. Coherent detection schemes require prior knowledge regarding the transmitted waveform; also receivers need to synchronize themselves in frequency and phase with the transmitter and demodulate the received signal to extract its main features. This may be a good approach and has a good performance, but it requires a complex design and extra overhead in processing.

On the other hand, the non-coherent detection schemes have a major advantage due to their simplicity. They do not require prior knowledge of primary users but perform only a general check for primary user activity levels. A major disadvantage is poor performance under low signal-to-noise ratio (SNR) levels. In [10] authors proposed to increase the efficiency of the sensing detection using a dynamic threshold selection based on matched filter.

In this chapter we propose a well-designed scheme, a neural network-based energy detection, to detect the presence of primary users taking into account the simplicity of the design to increase the probability of detection. This is achieved by deploying neural network into the sensing engine algorithm so we can measure the energy level in a more efficient way, on a larger scale that is outside of the traditional simulation environment but using software-defined radio to realize the impact of real-time performance.

17.3 System Architecture

In this section, we present the computational resources which have been used to access software-defined radio nodes using the open remote access to CORNET (Cognitive Radio Network) testbed which is dedicated to researchers for testing and developing real-time wireless communication experiments. We also discuss the open source software framework CRTS which is used to control and develop our system and finally the proposed cognitive cycle for the cognitive engine model.

17.3.1 Computational Resources: Software-Defined Radio

Many testbeds are deployed in different places around the world to enable wireless research and development such as ORBIT [11], NITlab [12], CORTEXLab [13], and VT-CORNET [14]. In our implementation, we rely on VT-CORNET testbed for development and validation where it provides an open access platform through which researchers can rapidly develop and test cognitive radio (CR) and dynamic spectrum access (DSA) experiments and applications on software defined radio (SDR) nodes. This is achieved by cost-effectiveness and reliable accessibility to CORNET testbed which allows testing and development of software radio experiments remotely over secure shell (SSH) protocol.

(VT-CORNET) [15, 16] is a collection of high-performance servers and software-defined radio architectures for next-generation wireless system. And as we have mentioned before, developers around the world are rapidly prototyping, extensively testing, and comparing cognitive radio techniques, software-defined radio components, and dynamic spectrum access technologies on a larger scale that is outside of the traditional simulation environment. CORNET is divided into three layers/planes: radio, network, and user planes. Our main concern is the hardware capabilities so we focused on the radio plane which is equipped by Universal Software Radio Peripheral (USRP2) and the connection back to their respective General Purpose Processor (GPP) servers.

A major advantage of the USRP2 over similar USB-connected SDR platforms is the gigabit Ethernet connection between the host (GPP server) and the radio head (USRP2), which allows higher throughput. The USRP2 sampling specifications (16-bit ADCs, 14-bit DACs) permit the transmission of 25 million I and Q ("in-phase" and "quadrature") samples per second between the USRP2 and the host server; this translates to a maximum radio frequency bandwidth of 25 MHz after up-conversion. From the user plane perspective, we can control the radio head (USRP2) by accessing the GPP server remotely using an SSH connection for each node accessed at once and then we run the API commands of the USRP Hardware Drive (UHD).

One of the key advantages of VT-CORNET is that the testbed is specially licensed to operate over a broad range of frequencies, which provides researchers the opportunity to conduct experiments on live spectrum regardless of the ISM bands, in the presence of real primary users. CORNET operates under the authorization of federal communication commission (FCC) [17], where permitted bands are between 150 and 3600 MHz within the Virginia building where the testbed is located.

All users are responsible for ensuring that their research is conducted within the limitations of the FCC license.

17.3.2 Software Architecture

In our system, we used CRTS software [6] which provides a flexible framework for over-the-air test and evaluation of cognitive radio and dynamic spectrum access experiments. We can rapidly define new testing scenarios while customizing the behavior of each node individually using a system scenario controller; scenario controller is responsible for controlling the scenario test parameters such as the network throughput as well as the operating parameters of the radio, e.g., its transmit power, center frequency, antenna type, or sampling rate, and also the waveform parameters, e.g., modulation type, fast Fourier transform (FFT) size, or filter type. It provides a flexible way to automate the performance and testing of the radios behaviors.

System node controller is powered by a scenario file to define the experiment's number of resources, e.g., how many nodes will participate in the experiments, radio local Internet Protocol (IP) address, the virtual IP address of the node it initially communicates with, the network traffic pattern (stream, burst, or Poisson), the initial configuration of the CR, what logs should be kept during the experiment, and most important the cognitive engine allocation for each radio node. The main advantage of CRTS is the flexibility to include user-defined embedded cognitive engine written in C/C++ or to include a GNU Radio python script to act as a cognitive engine or even to include the generated python files from GNUradio companion flowgraph to act as a software radio engine. We designed the primary user and secondary user engine using C/C++ and the spectrum analyzer using GNURadio uhd_fft python script.

17.3.2.1 Scenario Controller

Scenario controller is a part of the CRTS software that provides control the computational resources, SDR nodes, and manage the engine/application which will be ruined on each of them and how they interact with each other with user-defined parameters and execute it accordingly. This provides a user-friendly configurable way to initiate the connection, receive feedback, and exert control over a scenario's operation in real time. As shown in Fig. 17.1, CRTS controller controls and initiates the connection between two USRP nodes, and it is powered by a scenario configuration file which defines how many nodes are involved in our experiment and their configuration.

17.3.2.2 Experiment Scenario

As we have discussed before, the CRTS scenario/experiment controller will run the tests specified by a scenario master configuration file. The configuration file specifies the number of scenarios to be run, their names, and the number of times each scenario will be run which can be specified once for all scenarios, or for each individual scenario. In our case, we run only one scenario that contains four USRP nodes: spectrum analyzer, primary user, cognitive user, and scenario controller as illustrated in Fig. 17.2.

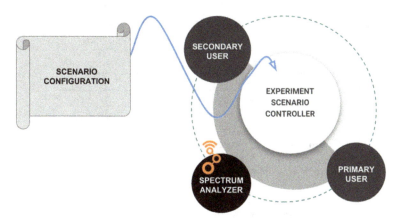

Fig. 17.1 CRTS scenario controller

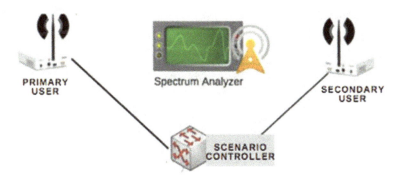

Fig. 17.2 SDR-based system architecture

The configuration scenario specifies the number of nodes in the experiment, which are three in our case hosted on the scenario controller makes them four in total, and the duration of the experiment. Furthermore, each node configuration parameters as mentioned before, local and virtual IP, cognitive engines, and the physical layer parameters.

17.3.3 Primary User's Engine

In this system, we developed two engines, one for primary users to be our reference and target and the other for secondary users which is designed to observe the primary user activity. The primary user (PU) will send a random stream data based on OFDM waveform with a 16 QAM modulation taking place, and then the data are transmitted over the air. To achieve the uncertainty/random activity, the primary user's engine is powered by *PU Random Behaviour* Engine which generates a random channel to operate on it [CHANNEL1, CHANNEL2, CHANNEL3] without any predetermined probabilities using the standard rand method for generating random integers and then divides by the maximum number that can be generated which is 3.

Algorithm 1 PU engine

---

1 Generate a random channel $(ch1, ch2, ch3)$

        Input : Three channels fc (center frequency)

        $ch1$ $ch2$ and $ch3$

  Output: *random index* holds the selected channel

2 Set transmitter $(random\ index)$;

  Input : *random index*

  Output: *operating freq*

  if (random index == ch1) **then**

    set operating freq to ch1

  else if (random index == ch2) **then**

    set operating freq to ch2

  else

    set operating freq to ch3

  end if;

  $TimerTic \leftarrow 0$;

  $StartTx \leftarrow opratingFreq$;

  while $TimerToc \geq SensingTime$ **do**

    $TimerTic \leftarrow 0$;

    GoTo 1

  end

17.3.4 Secondary User's Engine: Cognitive Engine

In Fig. 17.3, we illustrate a simplified cognitive cycle, where we start sensing the radio environment using energy detection method and then the sensory data are mapped into the neural network supervised learning model to find out the corresponding solution to be trained, so later it can predict the spectrum holes. Further secondary user radio can dynamically configure the corresponding parameters like transmitted power, sampling rate, center frequency, or the wave form. In our model, we focus on changing the center frequency to the channel where the primary user is absent.

In Algorithm 2, the secondary user is powered by neural network-based energy detection cognitive engine. The SU senses three channels simultaneously at the spectrum band 800 MHz at fc = 833e6 and B.W = 13e6, where fast Fourier transform (FFT) is applied and received energy of the observed channels is detected. We get 512 samples and do the discrete Fourier transform using FFT algorithm finally, we sum the FFT pins around our desired channels to get the signal strength of each channel; cognitive engine configuration is shown in Table 17.1. Also, it is important to know how many packets should be received according to the sensing time and USRP receiver rate; then the detected/extracted features could be mapped into the neural network classifier to make a prediction which indicates whether the channel is free or occupied; the main subroutine is illustrated in Fig. 17.4.

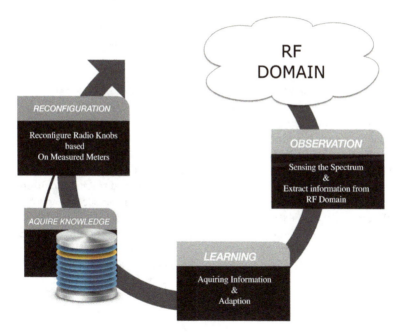

Fig. 17.3 Secondary user's cognitive cycle

Algorithm 2 SU Cognitive engine

---

1 Enable Sensing($Sampling Rate, Center Freq, FFT size$);

2 Features Extraction ($random\ index$);
Input : ($nchannels$)
Output: ($NF,\ Ech1,\ Ech2,\ Ech3$)
$i \leftarrow 1$;
while $i \leq nChannels$ **do**
  execute FFT
  $Ech(i) = \sum_{n=Fbins_l}^{Fbins_u} FFTBuffer(n)$
  $i \leftarrow i + 1$
end
NN Classsifier ($NF,\ Ech2,\ Ech3$);
Input : *Feature Vector ($NF, Ech1, Ech2, Ech3$)*
Output: *Channel Status*
    ($CH1status, CH2status, CH3status$)

Table 17.1 Cognitive engine physical layer configuration

Knobs	Configurations
Waveform	OFDM
Modulation	16 QAM
FFT length	512
Sampling rate	13 Msps
Sensing frequency	0.8 Hz
Sensing period	100 ms
Frequency band	800 MHz
UHD gain	20–29 dB

We followed in our systems two sensing approaches: a discrete sensing mechanism using the USRP2 and a continuous sensing mechanism using the USRPX310. In the first method, the sensing frequency might not necessarily affect detection performance but might affect timing performance and the throughput of the system. However, by using USRPX310 we can continuously sense the spectrum, where there are two daughter-boards so we can sense and transmit at the same time thus enhancing the throughput, but this costs an overloaded resources consumption from processing and power perspectives. Therefore, in the second approach the neural network is trained while the prescience of SU transmission with a waveform with a narrow bandwidth and low power, so in real time SU can perform continuous sensing and transmission concurrently, but computation overhead affects the communication link and bandwidth between the host workstation and the USRPX300.

17 Real-Time Spectrum Occupancy Prediction

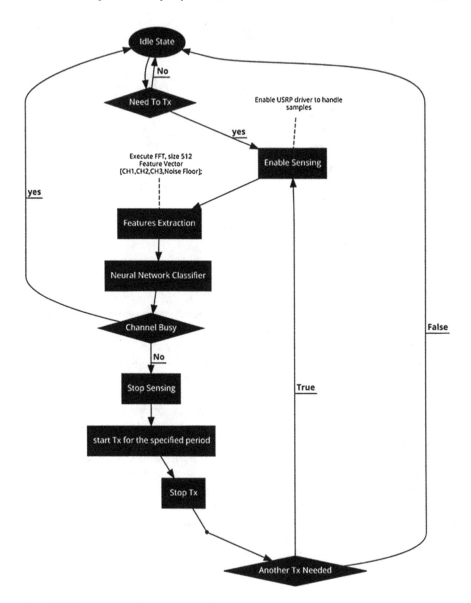

Fig. 17.4 Secondary user's cognitive engine state machine

17.3.5 Neural Network Configuration

Learning stage is a crucial component for cognitive radio to be self-aware of the environment and to achieve a highly embedded computational intelligence. We

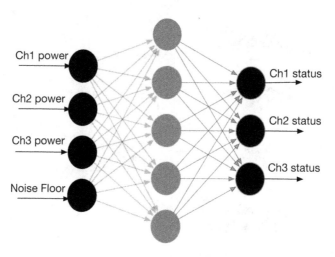

Fig. 17.5 Neural network topology

proposed the implementation of the brain inspired supervised algorithm neural network by designing a flexible network as shown in Fig. 17.5.

The neural network is capable of predicting the spectrum holes where the primary users are not active. The supervised training of the neural network was achieved by generating the data set, where we run a scenario by tuning one USRP to transmit on the channels CH1, CH2, CH3 or not transmitting at all with a fixed time interval, and the other USRP node is responsible for sensing the power of the three channels simultaneously plus the noise floor; finally, we train our model to adjust the neural network weights.

The general approach in learning involves presenting input training examples to an untrained network which is then used to get the output. The error is some scalar function of the weights that is minimized when the network outputs match the desired outputs. Then, the weights are adjusted to minimize the error so that the desired output could be reached. Our network is fed by the features extracted from the channel, and then at the output layer, we get the predicted channel vector which indicates the occupancy of desired channels. Neural network configuration is shown in Table 17.2.

17.4 Results

In this section, the performance of neural network-based energy detection is evaluated. We integrated our system members into a scenario involving the random activity of primary users and the cognition behavior of secondary users. First, we decide to follow the cognition cycle by observing and predicting the primary user activity as our first step and then taking action based on the predicted spectrum holes. The power level of both SU and PU is shown in Figs. 17.6, 17.7, and 17.8.

17 Real-Time Spectrum Occupancy Prediction

Table 17.2 Neural network configuration

Parameters	Configurations
Number of features	OFDM
Number of hidden layers	1
Number of hidden neurons	5
Number of output neurons	3
Activation function	Sigmoid
Training samples	500
Testing samples	200
Learning rate	0.001
Momentum	0.9
Number of epochs	80,000

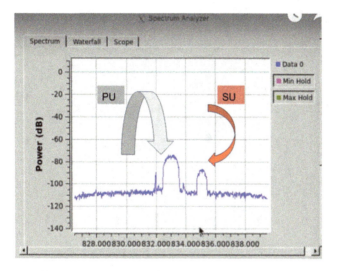

Fig. 17.6 Over-the-air coexistence PU and SU

In the table and confusion matrices below, we illustrate the probability of detection for the target channels. The sample data used for training and testing have been generated by tuning the USRP to send the data by tuning the transmitter gain which in turn targets and covers the desired range of SNR −30 to 20 dB. The result illustrates that the detection probability Pd can reach 100.0% for the SNR which is equal to or higher than −20 dB. The starting point of the implementation has been the original CRTS Code [18]. The developed code can be found in [19]

	SNR (dB)						
	−30	−20	−10	0	10	20	Avg accuracy
Prediction accuracy	96.0%	100.0%	100.0%	100.0%	100.0%	100.0%	99.3%

Fig. 17.7 Real-time terminal debugging

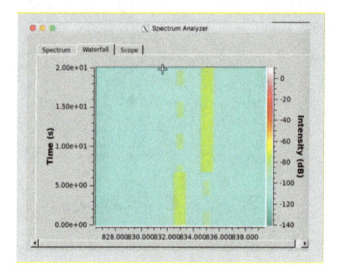

Fig. 17.8 Waterfall PU and SU coexistence

	CH1	CH2	CH3
CH1	0.95	1.0	1.0
CH2	1.0	0.96	1.0
CH3	1.0	1.0	0.96
−30 dB			

	CH1	CH2	CH3
CH1	1.0	1.0	1.0
CH2	1.0	1.0	1.0
CH3	1.0	1.0	1.0
−20 dB			

	CH1	CH2	CH3
CH1	1.0	1.0	1.0
CH2	1.0	1.0	1.0
CH3	1.0	1.0	1.0
−10 dB			

	CH1	CH2	CH3
CH1	1.0	1.0	1.0
CH2	1.0	1.0	1.0
CH3	1.0	1.0	1.0
20 dB			

17.5 Conclusion

This chapter proposes to test and verify the use of spectrum sensing techniques over the air outside the conventional simulation environment by deploying artificial neural network-based energy detection method. The method achieved perfect classification and detection outcomes which is based on comparing the received signal to an adaptive/dynamic threshold value. Unlike previous works, matched filter and cyclostationary feature detection methods require a perfect knowledge of the signal of interest, which adds processing overhead, but both have a good performance under low SNR. This chapter uses a simple scheme, neural network-based energy detection, which due to its simplicity and reliable performance, which achieves good detection outcome under low SNR proven by testing the system outside the conventional simulation environments using software-defined radio. In the future we intend to add a new methodology based on Markov model capable of selecting the optimum free channel based on primary users transitions is proposed.

We also are interested in tracking the patterns and behavior of primary users hoping that secondary users can exploit the channel and reduce computation overhead by switching off the sensing engine.

Acknowledgements We would like to thank Virginia Tech CORNET Testbed. This work would not have been possible without the open remote access to their computational resources.

References

1. J. Mitola, G. Maguire, IEEE Pers. Commun. **6**(4), 13–18 (1999). https://doi.org/10.1109/98. 788210
2. S.S. Haykin, *Cognitive Dynamic Systems: Perception-Action Cycle, Radar, and Radio* (Cambridge University Press, Cambridge, 2012)
3. P.T.V. Bhuvaneswari, in *Introduction to Cognitive Radio Networks and Applications* (CRC Press, Boca Raton, 2016), pp. 83–97. https://doi.org/10.1201/9781315367545-6
4. M. Sherman, A. Mody, R. Martinez, C. Rodriguez, R. Reddy, IEEE Commun. Mag. **46**(7), 72–79 (2008). https://doi.org/10.1109/mcom.2008.4557045
5. D.R. DePoy, *Cognitive Radio Network Testbed (Cornet): Design, Deployment, Administration and Examples*, Master's thesis, Virginia Polytechnic Institute and State University, 2012
6. E. Sollenberger, F. Romano, C. Dietrich, in *2015 IEEE 82nd Vehicular Technology Conference (VTC2015-Fall)* (2015). https://doi.org/10.1109/vtcfall.2015.7391168
7. T. Yucek, H. Arslan, IEEE Commun. Surv. Tutor. **11**(1), 116–130 (2009). https://doi.org/10. 1109/surv.2009.090109
8. F. Xu, X. Zheng, Z. Zhou, in *2009 9th International Symposium on Communications and Information Technology* (2009). https://doi.org/10.1109/iscit.2009.5341166
9. I.F. Akyildiz, W.Y. Lee, M.C. Vuran, S. Mohanty, Comput. Netw. **50**(13), 2127–2159 (2006). https://doi.org/10.1016/j.comnet.2006.05.001
10. R. Singh, S. Kansal, in *2016 IEEE Students Conference on Electrical, Electronics and Computer Science (SCEECS)* (2016). https://doi.org/10.1109/sceecs.2016.7509355
11. Open-access research testbed for next-generation wireless networks (orbit) (n.d.). http://www. orbit-lab.org

12. Network implementation testbed using open source platforms (n.d.). https://nitlab.inf.uth.gr
13. Fit/cortexlab, cognitive radio testbed (n.d.). http://www.cortexlab.fr
14. Cognitive radio network testbed (n.d.). https://cornet.wireless.vt.edu
15. T.R. Newman, A. He, J. Gaeddert, B. Hilburn, T. Bose, J.H. Reed, in *2009 5th International Conference on Testbeds and Research Infrastructures for the Development of Networks & Communities and Workshops* (2009). https://doi.org/10.1109/tridentcom.2009.4976217
16. T.R. Newman, T. Bose, in *2009 IEEE 13th Digital Signal Processing Workshop and 5th IEEE Signal Processing Education Workshop* (2009). https://doi.org/10.1109/dsp.2009.4786023
17. WF2XRP FCC license (2015). https://cornet.wireless.vt.edu/license.html
18. Cognitive radio test system (2015). https://github.com/ericps1/crts
19. Cognitive radio network (2017). https://github.com/astro7x/Cognitive-Radio-Network

Chapter 18
SCC-LBS: Secure Criss-Cross Location-Based Service in Logistics

Udai Pratap Rao, Gargi Baser, and Ruchika Gupta

18.1 Introduction

Transportation of goods from a starting location to a previously known end location is known as delivery. The study involving the transference and delivery of a variety of goods is popularly known as Logistics [18, 19]. Businesses specialize a variety of vehicles for distribution of the goods at the correct location. Logistics study dealing with physical items usually involves various components like correct information flow regarding handling of materials, packaging details, inventory, and many a times security also. Communication has to be established between these vehicles for effective delivery of goods. The vehicles must be able to communicate with each other being at different locations during transits. This leads to consideration of the moving vehicles being able to locate each other in large ad-hoc network. Such system is alluring since no pre-fixed infrastructure investment is required for communication. Instead, in these systems the mobile units form an autonomous association with each other voluntarily, after agreeing to transmit each other's packets towards their destination. Such wide range of moving vehicles will have continual topology changes and will also encounter several disconnections amongst themselves. The routing protocols that are based on topology or topology-based routing protocols have performed less effectively for such varying networks. Geographical routing protocols have helped in enhancing the performance and improving the scalability of such wide ad-hoc networks. These routing protocols

---

U. P. Rao (✉) · G. Baser
Computer Engineering Department, S V National Institute of Technology, Surat, Gujarat, India
e-mail: upr@coed.svnit.ac.in

R. Gupta
Computer Science Engineering Department, Chandigarh University, Chandigarh, India

© Springer Nature Switzerland AG 2019
I. Woungang, S. K. Dhurandher (eds.), *2nd International Conference on Wireless Intelligent and Distributed Environment for Communication*, Lecture Notes on Data Engineering and Communications Technologies 27,
https://doi.org/10.1007/978-3-030-11437-4_18

make the use of location-based services to gain information regarding the locality of destination node and also orientation of neighbors.

The delivery services must ensure that the data captured and exchanged by these vehicles specialized for delivery of goods must be secure. The moving vehicles have embedded sensor units that are responsible for the exchange of data. Since these sensor units include low priced resources, low capacity, and a lightweight platform, it is hard to implement the already existing cryptographic algorithms on these sensors. In this paper, we propose a scheme known as secure criss-cross location-based service (SCC-LBS) for the vehicles to communicate with each other, while delivering different types of goods (like fruits, clothes, cosmetics, home appliances). Such scheme is beneficial for wholesale distributors who perform delivery of goods at different places. Firstly, this scheme overcomes the drawbacks of existing fixed infrastructure communication network and scales over a large number of mobile nodes. Secondly, it is the sensor units embedded over moving vehicles that collect and disseminate crucial and sensitive data such as delivery statistics and change in plans once communication establishes. The data stored in sensor units is kept private and secured by applying lightweight operations on cryptographic algorithm suitable for sensors.

The rest of the paper is framed as follows: Sect. 18.2 reflects the related work. Section 18.3 contains features of our approach. Section 18.4 exhibits the system architecture and methodology of our proposed SCC-LBS approach. Performance results and analysis are emphasized in Sect. 18.6. Finally, we conclude our work in Sect. 18.7.

18.2 Related Work

Classification of LBS for ad-hoc networks can be done in the following ways: flooding based and rendezvous based. Figure 18.1 illustrates the various location-based services for ad-hoc network of vehicles. Reactive and proactive services compose the former category. Every node is required to flood information related to its position in the entire network at regular intervals in the proactive flooding-based location service. Thus, every node performs updating of its location table. DREAM [2] is the prominent example of such service. The basic concept of DREAM is that the far away node (the most distant node) seems to be moving slower as compared to its neighbor moving at the same speed. Hence there is a decrease in updating of frequency with the interspace to the node. However, highly mobile nodes need to send more packets regarding location upgradation.

In the reactive flooding-based location service, the response is conveyed only when a request regarding location is received. This ensures avoidance of overhead because of unusable location information received. High latency which is not appropriate for the vehicular network is the major drawback of such services. RLS [4] is an example of such type of service. The latter category known as the rendezvous-based location service involves conformation of all the nodes in the

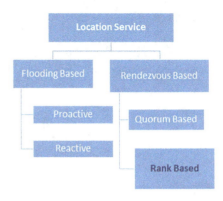

Fig. 18.1 Location-based service for ad-hoc networks

network for a distinctive mapping of a node with remaining nodes. The appointed nodes known as the "location servers" perform the dissemination of geographical location information. These services are composed of two operations:

- **Location Update:** Nodes in the network perform updating of their location through the specially recruited location servers.
- **Location Request:** In order to know another node's geographical location, a node has to send a request to the location servers, which will respond straightaway when they receive the request from a particular node.

Quorum-based approach [14, 30] is an example of rendezvous-based location service. This approach involves two types of clusters, namely update quorum and query quorum. Update quorum is the cluster of nodes that collects location updates sent from different nodes. Query quorum is another cluster of nodes that is responsible for answering location related queries. It's not mandatory to keep the mentioned two group *disjoint*. Authors in [14] discuss the major challenges regarding query generation process in such system. Protocols for geographical redirecting are required for exchanging messages between vehicles, in VANETs (vehicular ad-hoc networks). Geographical redirecting takes physical and network closeness into consideration. GPS [10] helps every node in the network to identify its location. Hello packets are periodically disseminated to neighbors by a node to notify its existence, velocity, and geographical position. Tables recording position and present neighbor's identity are maintained by each node. The header part of the packet targeted to a specific node includes identity of destination node and its geographical location. Neighbor table is consulted for choosing the nearest neighbor closest to a particular node say X, so that node can direct the packet with the help of the neighbor to X. The same procedure is followed by the chosen nearest neighbor till the packet reaches X. We introduce a new category (refer Fig. 18.1), the rank-based approach which will be utilized by SCC-LBS.

There also exist several distributed location services such as peer to peer spatial cloaking [5] that establishes a communication chain between several units. However, realization and establishment of an ad-hoc family of node experiences delay in this scheme.

Table 18.1 Different geographical redirecting location services

Approach	Characteristics	Drawbacks
Protocol for dynamic routing in mobile ad-hoc network [2]	Proactive scheme that involves flooding of location updates amongst all nodes	Not suitable for large networks, overhead occurs because of sending unwanted location data
Location aided routing [15]	Reactive scheme in which queries regarding location are answered only upon receiving a location request	Occurrence of high communication latencies, the least desired feature in vehicular environment, also not suitable for large networks
Finn's Cartesian routing [9]	Assigning special identifier to each node along with location information	Details regarding identifying node location and mobility management are not mentioned
Loop free method for routing [25]	Use of local information associated with each node for routing	Suitable only for unit graphs
Landmark system [27]	A hierarchy is maintained for routing provision in dynamic network	Vulnerable because of moving nodes near to the tip of hierarchical structure

Cartesian routing proposed by Finn [9] assigns a unique identifier to each node along with its geographical location. Scoped flooding of packets is done to handle dead ends. However, no details regarding handling of mobility and identification of node location are discussed. DREAM [2] technique involves flooding of position updates in the whole region, so that each and every node can revise and form database regarding positions. On the other hand, LAR [15] is a reactive scheme that performs flooding of position only when it is required to know the position of destination node. Both the schemes are not appropriate for large ad-hoc networks.

A scheme popularly known as the landmark system [27] ensures routing provisions for dynamic network. A hierarchy is maintained for routing provision. Nodes are assigned with IDs that are permanent along with a changeable landmark addresses. Routing can easily be performed using landmark address. The permanent IDs assigned to each node are linked to the present address of nodes for provision of actual location service. The major perks of this scheme are that it limits the storage space associated with each router. Moreover, landmarks can be easily elected dynamically using the scheme, hence making it suitable for large network areas. Table 18.1 summarizes characteristics and drawbacks of different geographical redirecting location services.

As mentioned above that sensor units are embedded on the moving vehicles that are responsible for communication between moving vehicles or nodes in the network, the sensor network formed plays a major role in distributed supply chain and logistics [1]. The continuous evolution of these networks has led to thrilling and strenuous research areas. The sensor unit on each vehicle is responsible for disseminating crucial and sensitive data. There are several resource constraints associated with wireless sensor networks, thus security models designed for ad-hoc network cannot be used for sensor networks.

While developing key management scheme for WSN several points need to be considered such as:

- **Battery life:** Battery life is limited in sensor units. Making use of asymmetric key management scheme like public key cryptography [22] is infeasible since such schemes consume more energy while performing intricate mathematical calculations. Symmetric techniques having less computational means (involving high complex mathematical operations) and consume less energy during computation should preferably be used.
- **Memory:** 6–8 Kbps of memory is usually available for sensor units, a part of which is engaged by the operating system such as TinyOS, associated with the network. The minimum storage left has to be now utilized productively for key storage or for caching received messages.
- **Prior deployment knowledge:** Sensor networks are randomized such that the sensor units are placed arbitrarily in a dynamic way. Hence, the maintenance of data regarding their placement becomes impractical. During initialization of keys for the key management scheme, the location of nodes should not be known.
- **Resistance:** Replication of nodes that are compromised by an adversary in the network can take place, once an adversary gains access to the data of few nodes in network. The whole network is then under the attack of the adversary because of deployment of these replicated nodes. Soon adversary gains control over the entire network. In order to avoid such attacks replication of nodes must be resisted.

Based on our study we present characteristics and limitations of key management schemes associated with wireless sensor networks (WSN). The simplest of all is the single network wide key [8]. In the first phase of this scheme, nodes in the entire network are preloaded with a single key. Second phase utilizes the key for encryption and decryption of messages in the network. This scheme requires least storage and avoids use of complex protocols. Since the key is preloaded the nodes require no sharing of keys and can easily send messages by utilizing the already shared key. The major drawback of this scheme lies in the fact that once a particular node gets compromised the entire network is at risk because of the shared key.

Pairwise key establishment scheme [17] provides resistance from node replication and node to node authentication. Pairwise keys are allotted for each node with every other node in the network. Unique pairwise sharing of keys between nodes ensures provision of node to node authentication. The disadvantage is the additional overhead incurred because of the $(n - 1)$ keys that each node has to establish for the remaining nodes of the network. From the memory point of view the maintenance of such large number of keys is costly.

It is observed that the computationally faster public key schemes [20, 24, 29] are not suitable for WSN. The typical public key schemes cannot consider the resource restrictions of sensor networks such as scanty memory, evaluation and measuring capacity, and limited power supply. Moreover these schemes involve expensive computation.

Trusted base station scheme [28] exchanges session keys amongst nodes for communication with the help of mediator called the "trusted base station." Popularly known as centralized key distribution center (KDC) scheme associates several scalability issues with it and the trusted base station becomes the only center of attack for adversaries. Hierarchical key management [7, 26, 31] requires partition of network into hierarchy. The data flows through this hierarchy and imposes a large amount of communication overhead while forming cluster hierarchy. Requirement of additional resources by the formed clusters enhances the overall cost of the network.

Authentication: u TESLA [21] provides strong security along with reduced cost for communication. However, more overhead is generated when keys are released in u TESLA, results in message delay, and obstructs communication of critical messages in real time application. Moreover authors in [11–13, 19] have discussed various schemes for achieving privacy in location-based service. Table 18.2 summarizes characteristics and drawbacks of abovementioned schemes. We have tried to overcome the limitations of the abovementioned schemes in SCC-LBS.

18.3 Features of Proposed Scheme

In secure criss-cross location-based service (SCC-LBS) we combine geographic redirecting and a technique to determine node's location to form a network like layer that allows a vehicle to send data packets to another vehicle in the network. In location-based services where there is a fixed location server, vehicles need to send periodic location updates to this fixed server. Vehicles then enquire this server to know the location of destination vehicle (or node) before utilizing geographic redirecting to send data. However, there are many problems associated with the use of one location server such as in case of single concentrated server, vehicles close to each other pick up no points of interest—they are supposed to contact conceivably far off area server with a specific end goal to impart data locally. Additionally schemes with single location server do not permit numerous system partitions to function regularly in their own segment. Moreover, issues related with reduced scalability to extensive number of mobile vehicles are highly probable under the centralized server techniques. With the specific end goal to spread work amongst all nodes, SCC-LBS also evades method such as progressive system to decide area server duty. Undue weight is put by such systems on nodes that are unfortunate enough to be chosen as a pioneer or those nodes that are set at upper heirs in the chain of command.

18 SCC-LBS: Secure Criss-Cross Location-Based Service in Logistics

Table 18.2 Various key management schemes for sensor units

Approach	Characteristics	Drawbacks
Single network wide key [8]	Nodes in the entire network are preloaded with a single key	Once a particular node gets compromised the entire network is at risk
Pairwise key establishment scheme [9, 15]	Pairwise keys are allotted for each node with every other node in the network	Additional overhead incurs due to the generation of $(n - 1)$ keys by each node
Public key schemes [1, 11, 12]	Involves two keys "public and private key" for security. Details of public key are shared with everyone whereas only the owner knows the details of key that is private	Expensive computation cost, not suitable for resource constraints of sensor networks like less memory, etc.
Trusted base station scheme [13]	Trusted base station responsible for the exchange of session keys amongst nodes	Trusted base station is prone to the single point of attack
Hierarchical key management [2, 14]	Requires partitioning of whole network into hierarchy	Formation of cluster hierarchy imposes large amount of overhead, increased cost of network due to requirement of more resources by the clusters
Authentication: u TESLA [16]	Provision of strong security along with reduced cost for communication	Not suitable for real time applications, overhead generated when keys are released, message delays may also occur

Lastly, when the communication is established between these vehicles in the network, it might not be secure.

To address these problems we propose a scheme called secure criss-cross location-based service (SCC-LBS). The characteristics of our scheme are as follows:

- SCC-LBS works adequately well for disconnected pockets of vehicles.
- It has no reliance over specific vehicles; therefore, it is a fault tolerant approach.
- SCC-LBS scales up to a larger quantity of vehicles, and likely gives an administration that scales to an extent of an expansive metropolitan region.
- Geographic redirecting allows each vehicle the capability to decide the location associated with any node it can reach. That implies an area query must not include the vehicles that are too far or "off the beaten path nodes."
- It regularizes the work of updating location amongst all vehicles.
- It enables routing in huge ad-hoc network that comprises of mobile units.

- The scheme ensures security of data interchanged between vehicles in the network at nodal and network level. Brute force attack, algebraic attack [6], and slide and related key attacks [23] are taken into consideration to evaluate the performance.

18.4 Secure Criss-Cross Location-Based Service

The proposed system SCC-LBS works in three phases viz. selecting location updaters, querying location updaters, and security of the data once the connection is established. We take into consideration that the enterprise is assigned the task of delivering several kinds of goods like fruits and vegetables, cosmetics, etc. The enterprise which has been given the task of delivery of goods has a specific number of vehicles (trucks, vans) for each of the category of goods. The specialized vehicles are numbered in alphanumeric order, for example for fruits and vegetables delivery the vehicle series will be P1,P2,P3... and so on, similarly for cosmetic delivery it will be Q1,Q2,Q3..., so on. A vehicle successfully identifies its own location using the global positioning system (GPS) [10] that is embedded on it. A vehicle **P** in SCC-LBS chooses an arrangement of area servers in a way such that the chosen arrangement is different from those that are chosen by other vehicles in the network. Additionally, the entering and leaving of vehicles in the network does not affect the choice of area servers for vehicles (or nodes). Vehicles scanning for **P** can find **P**'s area servers utilizing no earlier information past vehicle's **P**'s alphanumeric ID. This is refined via completing much a similar convention that **P** uses to choose its servers initially in the network. Note that we use term "node" to describe vehicles throughout the paper.

18.4.1 System Architecture

The area is divided in the form of a pattern that contains intersecting straight lines or paths as shown in Fig. 18.2.

Initially, a similar worldwide division of area is known to all the nodes in the network. The division is such that it forms a progressive system of networks that has squares of expanding size as shown in Fig. 18.2. A minor single square is alluded as rank 1 square shown using blue colored area in Fig. 18.2. A group of four rank 1 squares makes a rank 2 square and similarly the entire ranking takes place. It can be seen that the green colored region in Fig. 18.2 is not a rank 2 square; this states that not all squares of rank K forming a quad form a rank $(K + 1)$ square. This is to avoid co-occurrence, such that a rank K square will be associated with only a single order $(K + 1)$ square, not quartet (or four). Additionally, it states that a node shall accommodate into a single square of particular rank. The expanding arrangement of square sizes gives a setting in which a node chooses less area servers at more

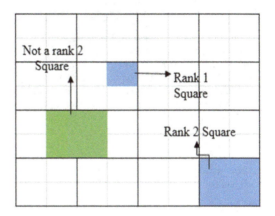

Fig. 18.2 Example of division of area

noteworthy separations. Notice that some other adjusted various leveled segment of the space can also be utilized for division of area.

Algorithm 1 represents the pseudocode for SCC-LBS algorithm.

18.4.2 Phase 1: Identifying Location Updaters

The main issue is how a vehicle selects its location updaters, by utilizing its alphanumeric ID and the foreordained hierarchy. A vehicle's technique is to enlist vehicles with alphanumeric IDs "nearby and greater" to its own particular ID to fill in as its location updaters so that different vehicles can find it. Suppose vehicles wish to contact **P19** and in order to do that the vehicle nearest to **P19** in ID space is to be the vehicle with the minimum ID more prominent than **P19**. This is used by the vehicles to select their area servers. There are few points to keep in mind while selecting location updaters:

- The ID space is considered to be roundabout, **P2** is nearer to **P19** than **P7** is to **P19**.
- A vehicle while selecting and querying a particular vehicle first checks for its alphabetical order and then considers numeric part of the ID. For example, we can see in Fig. 18.3 that **P25** is chosen as location updater by **P19** instead of **Q23** in spite of 23 being least ID bigger than 19.
- Each of the sibling square of the target square where the vehicle is present is looked upon to select location updater.

The correct points of interest of the determination are best comprehended with a case (described in Fig. 18.3). Three area servers are picked up by a node for each level in the partitioned area (rank 1, 2, etc.). For instance, in Fig. 18.3, **P19** initiates three updaters all together for rank 1 squares, three updaters all together for rank 2 squares, and all three updaters together also for rank 3 squares. The triad of the rank

Algorithm 1 Pseudocode for SCC-LBS algorithm

1.//**Phase 1: Selecting Location Updaters** 2. Let Vehicle XY wants to select its updaters (X is the chronological series assigned to vehicle & Y is the numeric ID)

3. Initially S(XY)=0 // S(XY) is set of vehicle's selected updaters

4. Begin Search in sibling Square at level n // initially n=1

5. While (vehicle numeric ID least>own numeric ID) //ID space considered to be circular

6. { Check for found vehicle if (X==chronological order of searching vehicle)

7. If yes increment S(XY)

8. Else

9. Goto step 5.

10. } return S(XY)

11. Update each in S(XY) with location.

12. Begin search in level (n+1)

13. Goto step 5.

14. // **Phase 2: Querying Location Updaters**

15. Let Vehicle XY wants to query Vehicle X'Y'

16. Begin search locally with P(XY) // P(XY) set of vehicles for which XY is the updater

17. If (numeric ID least >vehicle X'Y')

18. { forward data immediately

19. Else Goto step 17.}

20. Continue till X'Y' updater found.

21. // **Phase 3: Security of data forwarded after connection established**

22. Let XY wants to send PT to X'Y' // PT is the data to be sent

23. Begin

24. Encrypt using Key Q1 // Key Q1 is network shared key inbuilt in sensors at time of deployment

25. { PT—>Key Q1—>PT' }

26. Forward to X'Y'

27. At X'Y' decrypt using Key Q1

28. { PT'—>Key Q1—>PT

29. PT''=PT + new data // Update data if required

30. PT''—>Key Q1—>PT''' // Encrypt modified data with Key Q1

31. PT'''—>Key Q2—>PT'''' //Encrypt using Key Q2, Key Q2 is dynamically changing key

32. After time 'T'

33. Stop Key Q2 generation

34. Decrypt data using last value Key Q2

35. } Stop

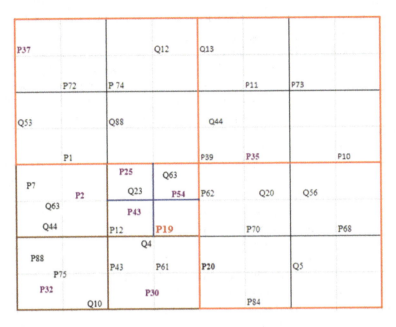

Fig. 18.3 Vehicle P19's location updaters

1 squares alongside **P19**'s personal's rank 1 square form a rank 2 square. **P19** picks the vehicle nearest to its ID space as an area server in each of these rank 1 squares. In the squares at higher level a similar process for choosing respective area servers can be observed. **P19** chooses **P2, P32,** and **P30** as area servers in the three rank 2 squares that join with **P19**'s rank 2 square to make a rank 3 square. The purple nodes represent the area servers of **P19**. At each tier the selection is as follows:

- Selection at rank1—P25, P54, P43 (blue bordered region in Fig. 18.3)
- Selection at rank 2—P2, P32, P30 (brown bordered region in Fig. 18.3)
- Selection at rank 3—P37, P20, P35 (red bordered region in Fig. 18.3)

As the node moves, location updaters of a node need to be updated with new position. To avoid extra traffic because of updates, nodes send updates relative to moved distance, i.e., rank 2 updaters are updated when node moves distance a, rank 3 when node moves to a distance $2a$, and so on. Generally, rank k updaters are updated by node after it has moved distance $2^{k-2} * a$.

18.4.3 Phase 2: Querying Location Updaters

To play out an area question, **P** sends a demand (utilizing geographic redirecting) to the slightest node (having nearest and larger ID) more noteworthy than or equivalent to **Q** for which **P** has position data. That node advances the enquiry in the same manner to other node, and the process repeats. In the end, the enquiry reaches a location updater of **Q** that forwards the enquiry to **Q** itself. Since the question contains **P**'s area, **Q** can react specifically utilizing geographic redirecting. The inquiry for location is sent to **Q** so **Q** can react with its most recent position. Note that we disregard an essential bootstrapping issue in which it is expected that nodes choose their location updaters fittingly and send them their coordinates appropriately.

At the point when there is no movement of nodes, the quantity of query advances that an area inquiry from **P** to **Q** takes is close to the request of the smallest square in which **P** and **Q** are co-located. Every area inquiry step probably requires a few geographic redirecting jumps. In Fig. 18.4, the whole graph is a rank 4 square. In this manner all inquiries can be performed in close to four area inquiry steps. At each progression, a question advances towards the best (nearest to the goal node in ID space) node at progressively larger amounts in the lattice order. In the beginning, the best node in the nearby rank 1 square receives the inquiry utilizing the neighborhood directing convention. Now with every step query moves towards bigger rank square. This is query's next move to the end goal node. This conduct not just confines the quantity of steps expected to fulfill an inquiry, it likewise limits the geographic district in which the question engenders. Since the question continues into bigger and bigger squares that however incorporate the source, the inquiry remains part of the smallest square that contains the source and the goal node (or destination).

To comprehend why each progression conveys the inquiry to the node that is ideal in the bigger square, we consider an example shown in Fig. 18.4 of a query from node **P78** for the location of **P19** (beginning of the process is indicated at the bottom right). The inquiry inconsequentially starts at the ideal hub, itself, in its rank 1 square since our shortened topology is close to one node for every square. The question advances to the ideal node **P21** in **P78**'s rank 2 square, on the grounds that **P78** is aware of the situations for all nodes that are in its rank 2 square.

Why **P21** is aware of the area of the ideal node in the following higher request square? Since **P21** is the ideal square in its rank 2 square we determine that the IDs of no nodes in that square have the vicinity of **P19** and **P21**. Presently, consider a node **P** which is not in 21's rank 2 square but however is placed in node **P21**'s rank 3. Review that an area server in node **P21**'s rank 2 square had to be picked up by node **P**. On the off chance that **P**'s ID is in the vicinity of **P19** and **P21** then **P** more likely will choose node 21 as an area server because rank 2 square of node **P21** has no better node that can be selected. In this manner, all nodes in its rank 3 square that lie in the vicinity of **P19** and itself are thought about for selection by node **P21**, including the least such node. For this situation, it is **P20**. Node **P20** must think about all nodes in the rank 4 square in the vicinity of **P19** and itself at subsequent stages. Since nodes **P20** and **P19** share a similar rank 4 square (the whole Fig. 18.4), node **P20** thinks about node **P19**, and the inquiry is finished. Tables 18.3 and 18.4 represent the contents of update and query packets in SCC-LBS.

18 SCC-LBS: Secure Criss-Cross Location-Based Service in Logistics

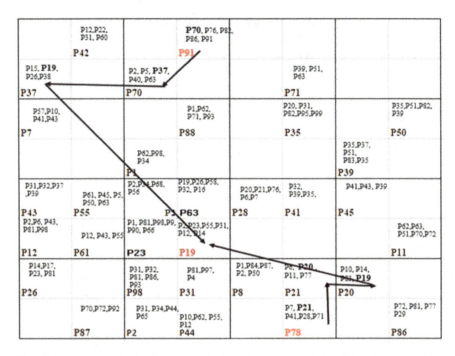

Fig. 18.4 A Possible location server organization of entire network (possible query from P91 to P19 and P78 to P19)

Table 18.3 SCC-LBS packet for updating location

Packet for updating location
ID of source
Location of source
Timestamp associated with source
Rank of destination point
Timeout for update
ID of next location server
Location of next server

Table 18.4 SCC-LBS packet for query

Packet for query
ID of source
Location of source
ID of target node
ID of next location server
Location of next server
Timestamp from last server's database

Error Handling There are two possible reasons for query failures: Initially, a node might get an inquiry bundle for **Q**, and is not aware of the area of any other node with an ID nearer to **Q** than itself. Such type of disappointment is moderately unusual. This will happen when an area server does not receive an area refresh for a node it shall think about. Since node's past refresh has been coordinated out by the server, the server has no means left to redirect the inquiry parcel. Several approaches exist to lighten such kind of disappointments, for example, utilizing stale area information in a final desperate attempt to redirect an inquiry parcel if the question would somehow fall flat. The second kind of query disappointment happens when an area server advances a bundle to the following nearest node's square, yet the node has moved out of that square (that is, the area data at the past area server is outdated). Since this disappointment mode is more typical, SCC-LBS contains specific means to reduce the issue.

Considering a node **P** which has moved out recently from the rank 1 square *t1* to another rank 1 square *t2*. Area servers of node **P**, especially the ones that are at larger distance, feel that **P** still exists in *t1* until the point when **P**'s next updates are received by them. **P** leaves a "redirecting pointer" in *t1* showing that it has relocated to *t2*. At the point when a parcel touches base in *t1* for **P**, the redirecting pointer ensures that the parcel is sent accurately. The redirecting pointer of **P** is communicated to all nodes in *t1* by node **P** before it departs from *t1*. In other words, the redirecting pointers can be thought of as being situated in the square *t1* instead of at a specific node. Along these lines, the redirecting pointers related with *t1* are sent to all nodes that move into *t1*, and once the nodes leave *t1*, they ought to overlook the redirecting pointers. A randomly picked redirecting pointer's subset that is saved at a node that is piggybacked with the nodes intermittent hello messages to proliferate redirecting pointers amongst all nodes in the rank 1 square and ensure that all new nodes entering the square are kept refreshed. After a hello message (consists of ID of source, its position and speed, record of IDs of its one hop neighbors, and the redirecting pointers) is heard, each redirecting pointer is included by node in the hello message, along with its own gathering of redirecting pointers, in case the pointer's unique supporter is in an indistinguishable square from the node. Along this path, redirecting pointer data is viably and spread effectively to each node of the square. Through this spreading technique, every node in the network that initially got **P**'s redirecting pointer is able to access data in the square regardless of whether it were to leave the square.

18.4.4 Phase 3: Security of SCC-LBS

In SCC-LBS scheme we have assumed that there are sensor units on the moving vehicles that are responsible for storing the exchanged information such as location information, redirecting pointers, or related data. In the given proposition the information display in every sensor embedded node gets separated into two sections:

Fig. 18.5 Various keys in SCC-LBS

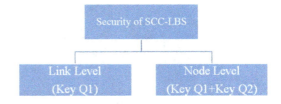

- **Setup data:** The codes of the projects and other information installed into the sensor's arrangement previously, which are utilized to process the detected data and to perform other activity identified with information transmission, security, and control. This setup data is impermanent.
- **Monitored data:** The data assembled by the sensor hub from its surroundings. We utilize two keys Q1 and Q2 for security of the data stored and exchanged between sensor nodes. As shown in Fig. 18.5 Q1 provides link level as well as node level security whereas Q2 is responsible only for node level security.

The process starts by inserting the solitary network wide shared key Q1 in all the sensors on vehicles, say Q1 goes under the classification of setup data. Q1 is a static key whose value never changes and shall be utilized for message exchange utilizing a symmetric encryption convention, an ultra-light weight cipher (PRESENT [3]) intended for WSNs. All sensor units likewise contain the data they have sensed from surroundings. The sensor units encode this information when it is assembled from nature utilizing the key Q1 and the cryptographic convention [3] utilized as a part of the message transmission. Once the encryption is done using Q1, message is transmitted to another node, hence encryption using Q1 and PRESENT [3] ensures link and node level security. Each node also contains a moment key Q2—called a varying key utilized for encoding the encrypted detected information along with the system wide shared key Q1, utilizing straightforward sensible, invertible operation (for example, XOR) with the goal that calculation does not become large.

A creative aggressor, for example, an assailant furnished with a workstation can without much of a stretch easily figure the key Q2, thus further uncovering key Q1. She then accesses the detected data that was encoded utilizing key Q1. The whole system gets bargained since Q1 is the main key that is in the shared system, for the purpose of exchanging messages amongst the distinctive sensor nodes. To resolve this issue the security risk value of Q2 is changed at specific intervals, thus making it a varying key. Whenever the aggressor verges on speculating the estimation of Q2, it gets change thus rendering Q2 uncertain for the adversary. In any case, before changing the estimation of Q2 the information and Q1 are decoded, both of which were scrambled by the older estimation of Q2, and again are encoded with the new estimation of key Q2. A protected time gap is kept to guarantee that the information and Q1 are sheltered amid the period for which they are left uncovered.

Detailed analysis for safe gap calculation is as follows: we have assumed key Q2 is of 64 bit value. Therefore, number of conceivable actions of $Q2 = T = 2^{64}*(1\div2) = 1.84*10^{19}*(1\div2)$. T likewise signifies the quantity of speculations

Fig. 18.6 Process of security in SCC-LBS

required by the aggressor to figure the estimation of Q1 utilizing brute force. In this way the likelihood that the adversary infers key Q2 = Guess = $1 \div (1.84*10^{\wedge}19 *(1 \div 2))$. Accordingly it is seen that the likelihood of speculating the estimation of Q2 is very less. Subsequently, if the adversary requires some investment "m" to perform one figure activity, at that point the time "n" required to perform T surmises = $1.84*10^{\wedge}19*(1 \div 2)*m$.

In this way, the perfect time after which key Q2 ought to be changed is n. However, to keep up the time gap as specified in the previous area to protect the information and Q1 (amid the immaterial period when Q2 is revised and the information and Q1 are re-scrambled), the time of revising the value of Q2 will basically be $n - \Delta$, where Δ refers to the time taken to re-scramble information and key Q1. Figure 18.6 presents the process of security in SCC-LBS.

18.5 Simulation Scenario and Evaluation Metrics

The simulations for the first two phases of SCC-LBS are performed using Ns-2. Greedy perimeter stateless routing (GPSR) [16] is the protocol we utilize for exchanging packets between nodes. We choose an area of 2500 m × 2500 m that represents part of Surat city. 802.11b Radio is the MAC layer used and transmission range is 300 m. Every simulation keeps running for 250 s. The nodes are put at consistently arbitrary areas. The simulation span is kept up a normal node density of around 50 nodes for each square kilometer. Purpose behind this decision is the anticipation that the framework to be utilized for moderately vast territories, for example, a ground or city, as opposed to in concentrated areas like a gathering lobby. Another reason is our anticipation that any sent framework will utilize radios permitting diminished power levels in regions with high hub thickness. The highest simulation we have performed is with 350 nodes (Table 18.5).

Every node moves utilizing an "arbitrary way-point" display. The node picks an irregular goal and pushes data towards it with a speed picked consistently in the vicinity of 0 and a most extreme speed of 20 m/s. At the point when the node achieves the goal, picks another goal and starts advancing towards it quickly (we have taken pause time as 50 s). Figure 18.7 represents the workflow of our simulation.

Table 18.5 Parameters for simulation

Parameter	Value
Tool used	Ns-2
Time	250 s
Number of nodes	50–350 nodes
Routing protocol	GPSR
Traffic model	Random waypoint
Transmission range	300 m
Terrain area	2500 m × 2500 m
Mobility speed	0–20 m/s
Packet length	64 bytes

Fig. 18.7 Workflow of simulation on Ns-2

We preferred Java programming for simulating the security phase. A client server architecture with socket programming is constructed to check adversary attack and energy consumption. The simulation is performed on Unix operating system running on Intel core i5 system having 64-bit architecture. Nodal level simulation is done using socket programming to check the strength of the proposed technique (use of keys Q1 and Q2 each of them being 64 bit keys along with simple operations) in shielding a solitary node from an adversary. The varying key Q2 is generated by random number generator that is devised using actual time of system. 50,000 μs is the interim at which system time obtained. Amid this interim the framework is sent to rest state to ensure energy saving.

The socket programming scenario involves two units: server node that acts as sensor and attacker node. At the server node, generation of keys Q1, Q2 takes place using HIGHT algorithm and random number generator respectively. On the other hand, the encrypted data using keys Q1 and Q2 is continuously sent at the attacker node, where continuous decoding of data by brute force technique takes place. The attacker node displays continuous results regarding whether the adversary has guessed the correct value for key Q2.

We use following parameters as performance evaluation metrics for comparison between secure criss-cross location-based service (SCC-LBS) and reactive location service (RLS).

Query success rate: This is interpreted as the proportion of the inquiries replied from every one of those replied with substantial location data (distance transmission scope of 350 m is permitted here).

Location-based service overhead: We have additionally contemplated the overhead prompted by the location-based administrations. The quantity of parcels reciprocated amid the area updates, inquiries, and answers are defined in terms of overhead.

Great scaling implies no rise in the measure of work performed by every node as an element of the aggregate number of nodes. We utilize the quantity of area database items every node has to store as measurement for our work.

The work required is expanded by movement in two ways. Initially, area servers need to be refreshed once a node moves. Secondly, if a node moves out of network, a few nodes may hold obsolete area data for it that makes inquiries for the node that moved more distant than expected. Taking care of versatility requires a trade-off between the transfer speed utilized by area refreshes and the transmission capacity accessible for information. In the event that a moving node sends refreshes forcefully, different nodes probably have the capacity to find it. In any case, the updates expend transmission capacity in rivalry with information. More regrettable, an exceptionally forceful refresh arrangement may cause dropping of updates because of network clogging. A node could send refreshes occasionally notwithstanding while moving rapidly, expanding the measure of transfer speed accessible to information. In any case, the data transfer capacity isn't helpful if the achievement rate of area inquiry turns out to be low, a direct result of wrong area data. The simulation demonstrates that SCC-LBS can accomplish a sensible trade-off for the decision of refresh rate. The energy consumption of the sensors can be given as:

$$\mathbf{P(Power)} = \mathbf{V(Voltage)} * \mathbf{I(Current)} \qquad (18.1)$$

where we consider: Voltage $= 4.5$ V and Current $= 1$ mA (suitable for sensor nodes). Therefore value of power will be $4.5 * 1 * 10^{\wedge}-3$ W.

We know that

$$\mathbf{Power(P)} = \mathbf{E(Energy)}/\mathbf{T(Time)} \qquad (18.2)$$

$$\mathbf{E} = \mathbf{P(power)} * \mathbf{T(Time)Joules} \qquad (18.3)$$

So energy consumption for our system can be calculated using above formula. Here "T" is the time taken to generate new value of Q2.

18.6 Results and Analysis

We compare our approach (SCC-LBS) with existing reactive location-based approach [4]. We provide statistical performance results and analysis of both the abovementioned approaches based on location-based service overhead and query success rate.

The threshold distance is 300 m and 802.11b radio bandwidth is 2 mega-bits per second. We assume the following points: A single node can be the starting point source for only one interconnection and it cannot be the end point of more than four interconnections. The creation of information traffic is by various steady piece

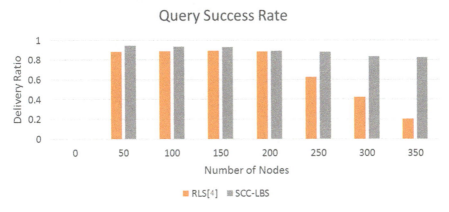

Fig. 18.8 Query success rate

rate associations is equivalent to a large portion of the quantity of nodes. With each interconnection formed, 64 byte packets are supplied per second for around 30 s.

Figure 18.8 shows the number of information parcels that are effectively conveyed. The greater part of the information parcels that SCC-LBS neglects to convey are because of SCC-LBS query disappointments; these parcels never leave the source. When SCC-LBS discovers the region of goal node, information misfortunes are improbable, since geographical redirecting adjusts effectively with the movement of middle of the road nodes. For around 200 nodes and less, a large portion of the message failures in RLS are because of broken source courses; at 250 nodes or more, failures are for the most part because of flooding-instigated clog. For RLS [4] the rate of successful query is least since the reply takes much more time to cover the entire network. Criss-cross completes a superior employment than RLS over the entire scope of quantities of nodes, particularly for substantial systems.

Figure 18.9 demonstrates the message overhead of the SCC-LBS and RLS [4] conventions. RLS does not make use of refreshing mechanism but it continuously sends packet in the network. On account of grid, the packets are HELLO, SCC-LBS refresh, and SCC-LBS inquiry and answer parcels. On account of LSR, the packets redirected are packet of request, response, stored answer bundles, and so on. SCC-LBS performs better in case of large ad-hoc networks whereas it is visible in Fig. 18.8 that RLS creates less convention overhead for small systems.

It is observed that at 200 hubs or more, network blockage takes place in RLS. This network blockage leads to dropping of half of the responses in network and stored responses messages, which further makes RLS [4] infuse considerably more request queries into the system. Additionally, as the system gets bigger, due to clog development the vulnerability of route associated with source increases. This leads to breakdown of network followed by RLS node source to send more request queries for parcels. RLS's overhead drops at 350 nodes since it is unable to send substantially more parcels within the sight of clog.

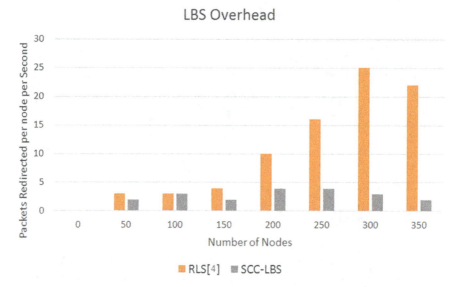

Fig. 18.9 LBS overhead

Fig. 18.10 Per-node database size (maximum)

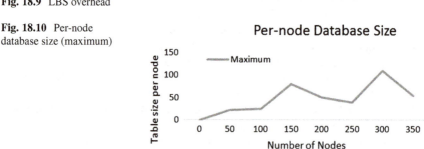

Figure 18.10 demonstrates the impact of the aggregate number of nodes on the extent of every node's SCC-LBS area table. The maximum area table size for all the nodes is incorporated in the plots. The peaks at 150 and 300 nodes in Fig. 18.10 are observed because simulated region does not precisely fill the levels in area, causing the load on database to be circulated unevenly. For such points, the database estimate is bigger in light of the fact that the squares that stretch out over the edge of the mimicked territory contain moderate number of nodes. Then again, the table size becomes normal gradually with the system estimate.

Our designed client server architecture for security analysis is as follows:

Server Node: At the server node, key Q2 is being generated dynamically. Moreover with the value of key Q2, XOR operation is being performed on the data that is already encrypted using key Q1 and PRESENT. Time mentioned is the safe gap that is kept up while decrypting the information with past estimation of Q2, i.e., before new Q2 is randomly generated. We claim that we provide node level security for each sensor unit for the calculated safe gap duration. The attacker node receives the encrypted data continuously, it is when the attacker utilizes brute force approach to guess the value of key Q2. We implement this scenario with the use of socket programming.

Attacker Node: The attacker node is where continuous guessing of Q2 takes place by the attacker. The simulation is done for a stretch of 1 day and it is observed that the attacker could not be able to identify the correct value of Q2. Hence, we safeguard our scheme from brute force attack. The algebraic attack requires deriving of system that is over defined with algebraic equations. The degree 8 round function of PRESENT (as per Boolean vector function) is impossible to be converted into an over defined system of equations hence, the algebraic attack is not possible.

In the slide and related key attack adversary utilizes key schedule weakness for attacking. Though our schemes sub-key generation algorithm is simple and linear it is resistant from attacks that occur because of round function utilization, along with avalanche effect and non-linearity [23]. It is realized that an iterated cipher with indistinguishable round functions is powerless against slide and related key attack. A unique constant for each round is utilized in PRESENT which shields it from slide attacks. We can see that the energy consumption is least because of use of lightweight operations like XOR. The amount of energy as calculated (around $0.12\,\mu J$) is suitable for sensor node.

18.7 Conclusion and Future Work

In this paper, we have proposed secure criss-cross location-based service (SCC-LBS) to establish communication between sensors embedded vehicles for delivery in goods, in a larger area. Proposed scheme incurs lesser LBS overhead for large sized network as compared to reactive location-based service. Our scheme SCC-LBS works adequately well for the disconnected pockets of nodes and eliminates the reliance on specific nodes, hence it is fault tolerant. SCC-LBS utilizes geographical redirecting which allows each node ought to have the capacity to decide the area of any node that it can reach. It stabilizes the work of location server amongst all nodes. Moreover the assigning of alphanumeric ID to nodes ensures that communication is established between similar categories of nodes. The scheme ensures security of data interchanged between nodes in the network, at nodal and network level. The following features can be considered as future work.

Security: We have utilized only single round of encoding data with logical operation XOR, but this might not give adequate insurance against the cutting edge cryptanalytic assaults; **Choice of Division of Area:** Rather than dividing the entire area into parallel lines, one might use other techniques for dividing the area (preferably the techniques that provide better scaling); **Mobility model:** Node movement can be accurately predicted by making use of suitable mobility models. Since our proposed scheme is random in nature, little work to speculate the speed of nodes in the network has been done.

References

1. Anjali, Shikha, M. Sharma, Wireless sensor networks: routing protocols and security issues, in *Fifth International Conference on Computing, Communications and Networking Technologies (ICCCNT)* (2014), pp. 1–5. https://doi.org/10.1109/ICCCNT.2014.6962992
2. S. Basagni, I. Chlamtac, V.R. Syrotiuk, B.A. Woodward, A distance routing effect algorithm for mobility (dream), in *Proceedings of the 4th Annual ACM/IEEE International Conference on Mobile Computing and Networking* (ACM, New York, 1998), pp. 76–84
3. A. Bogdanov, L.R. Knudsen, G. Leander, C. Paar, A. Poschmann, M.J. Robshaw, Y. Seurin, C. Vikkelsoe, Present: an ultra-lightweight block cipher, in *International Workshop on Cryptographic Hardware and Embedded Systems* (Springer, Berlin, 2007), pp. 450–466
4. T. Camp, J. Boleng, L. Wilcox, Location information services in mobile ad hoc networks, in *IEEE International Conference on Communications, 2002. ICC 2002*, vol. 5 (IEEE, Piscataway, 2002), pp. 3318–3324
5. C.Y. Chow, M.F. Mokbel, X. Liu, A peer-to-peer spatial cloaking algorithm for anonymous location-based service, in *Proceedings of the 14th Annual ACM International Symposium on Advances in Geographic Information Systems* (ACM, New York, 2006), pp. 171–178
6. N.T. Courtois, General principles of algebraic attacks and new design criteria for cipher components, in *International Conference on Advanced Encryption Standard* (Springer, Berlin, 2004), pp. 67–83
7. X. Du, Y. Xiao, M. Guizani, H.H. Chen, An effective key management scheme for heterogeneous sensor networks. Ad Hoc Netw. **5**(1), 24–34 (2007)
8. L. Eschenauer, V.D. Gligor, A key-management scheme for distributed sensor networks, in *Proceedings of the 9th ACM Conference on Computer and Communications Security* (ACM, New York, 2002), pp. 41–47
9. G.G. Finn, Routing and addressing problems in large metropolitan-scale internetworks. ISI Research Report (1987)
10. Global positioning system. https://en.wikipedia.org/wiki/Global_Positioning_System. Accessed 15 Oct 2017
11. R. Gupta, U.P. Rao, Achieving location privacy through cast in location based services. J. Commun. Netw. **19**(3), 239–249 (2017)
12. R. Gupta, U.P. Rao, An exploration to location based service and its privacy preserving techniques: a survey. Wirel. Pers. Commun. **96**(2), 1973–2007 (2017)
13. R. Gupta, U.P. Rao, A hybrid location privacy solution for mobile LBS. Mob. Inf. Syst. **2017**, 1–11 (2017)
14. Z.J. Haas, B. Liang, Ad hoc mobility management with uniform quorum systems. IEEE/ACM Trans. Netw. (TON) **7**(2), 228–240 (1999)
15. V. Hnatyshin, M. Ahmed, R. Cocco, D. Urbano, A comparative study of location aided routing protocols for MANET, in *IFIP Wireless Days (WD), 2011* (IEEE, Piscataway, 2011), pp. 1–3
16. B. Karp, H.T. Kung, Gpsr: Greedy perimeter stateless routing for wireless networks, in *Proceedings of the 6th Annual International Conference on Mobile Computing and Networking* (ACM, New York, 2000), pp. 243–254
17. D. Liu, P. Ning, Establishing pairwise keys in distributed sensor networks, in *Proceedings of the 10th ACM Conference on Computer and Communications Security* (ACM, New York, 2003), pp. 52–61
18. Logistics. https://en.wikipedia.org/wiki/Logistics. Accessed 23 Sept 2017
19. Logistics and security. https://www.penskelogistics.com/technology/secure-data/. Accessed 20 Nov 2017
20. D.J. Malan, M. Welsh, M.D. Smith, A public-key infrastructure for key distribution in TinyOS based on elliptic curve cryptography, in *First Annual IEEE Communications Society Conference on Sensor and Ad Hoc Communications and Networks, 2004. IEEE SECON 2004* (IEEE, Piscataway, 2004), pp. 71–80

21. A. Perrig, R. Szewczyk, J.D. Tygar, V. Wen, D.E. Culler, Spins: Security protocols for sensor networks. Wirel. Netw. **8**(5), 521–534 (2002)
22. Public key cryptography. https://en.wikipedia.org/wiki/Public-key_cryptography. Accessed 01 Nov 2017
23. M. Pudovkina, A related-key attack on block ciphers with weak recurrent key schedules, in *International Symposium on Foundations and Practice of Security* (Springer, Berlin, 2011), pp. 90–101
24. RSA. https://en.wikipedia.org/wiki/RSA_(cryptosystem). Accessed 15 Oct 2017
25. E. Takimoto, S. Aketa, Y. Otsuki, S. Saito, K. Mouri, A hybrid loop-free routing protocol for wireless mesh networks, in *International Conference on Frontiers of Communications, Networks and Applications* (ICFCNA 2014-Malaysia) (2014)
26. P. Traynor, H. Choi, G. Cao, S. Zhu, T. La Porta, Establishing pair-wise keys in heterogeneous sensor networks, in *Proceedings of 25th IEEE International Conference on Computer Communications*. INFOCOM 2006. Citeseer (2006), pp. 1–12
27. P.F. Tsuchiya, The landmark hierarchy: a new hierarchy for routing in very large networks, in *ACM SIGCOMM Computer Communication Review*, vol. 18 (ACM, New York, 1988), pp. 35–42
28. P. Vijayakumar, S. Bose, A. Kannan, Centralized key distribution protocol using the greatest common divisor method. Comput. Math. Appl. **65**(9), 1360–1368 (2013)
29. A.S. Wander, N. Gura, H. Eberle, V. Gupta, S.C. Shantz, Energy analysis of public-key cryptography for wireless sensor networks, in *Third IEEE International Conference on Pervasive Computing and Communications, 2005*. PerCom 2005 (IEEE, Piscataway, 2005), pp. 324–328
30. G.J. Yu, A quorum-based route cache maintenance protocol for mobile ad-hoc networks, in *Proceedings of the 22nd International Conference on Advanced Information Networking and Applications* (IEEE Computer Society, Washington, 2008), pp. 196–203
31. S. Zhu, S. Setia, S. Jajodia, Leap: efficient security mechanisms for large-scale distributed sensor networks, in *Proceedings of the 10th ACM Conference on Computer and Communications Security* (ACM, New York, 2003), pp. 62–72

Index

A

Additive random sampling (ARS), 187
Additive white Gaussian noise (AWGN), 146, 147, 184
Ad hoc on-demand vector (AODV), 194, 196
AES-NI instructions, 163
A* search-based next hop selection routing protocol (A*OR)
  delivery predictability, 62
  description, 62
  ONE simulator, 64
Automatic modulation recognition (AMR), 205
Average remaining energy, 41
Avoiding mistaken transmission table (AMTT) scheme, 195
AWGN, *see* Additive white Gaussian noise

B

Binary exponential backoff, 196
BitShift transformation, 162–164
Boyer Moore (BM) algorithm, 76
BUBBLE Rap protocol, 38

C

CESIS, *see* Cost-effective and self-regulating irrigation system
Chow's construction, 156, 157
Cognitive engine (CE), 220

Cognitive radio (CR), 183, 184, 204, 219
Cognitive radio test system (CRTS) software, 222
Collaborative-based schema, 28
Component-based software development (CBSD), 80–81
Concept map (CM), 83
Concept tree (CT), 82
CORNET testbed, 220
Cost-effective and self-regulating irrigation system (CESIS), 179
  activity diagram, 171, 172
  block diagram, 169
  class diagram, 171–173
  implementation
    cloud service provider, 175, 176
    cost of, 176, 179
    data retrieval, 175, 177
    Google assistant, 176, 179
    live data, 175, 176
    mobile application, 175, 178
    web page, 175, 177
  limitations, 167–168
  MQTT protocol and Wifi module, 168
  sequence diagram, 170–171
  system model, 173–174
  use case diagram, 169, 170
  variant of, 174–175
  water irrigation, 167
Covariance matrix theory (CMT), 184
Cyclic redundancy check (CRC), 19–20

© Springer Nature Switzerland AG 2019
I. Woungang, S. K. Dhurandher (eds.), *2nd International Conference on Wireless Intelligent and Distributed Environment for Communication*, Lecture Notes on Data Engineering and Communications Technologies 27,
https://doi.org/10.1007/978-3-030-11437-4

D

Data loss prevention (DLP)
- BM algorithm, 76
- concept map, 83
- content monitoring scheme, 77
- data loss modes, 79
- dataset, 87–88
- DCT, 83–84
- domain-specific ontologies, 77
- DR, 89–90
- DSS, 77, 84, 96, 97
- FDR, 79
- FIBO, 88, 94–95
- FNR, 79
- FPR, 79, 88–90
- frequency similarity metric, 90
- goal of, 77
- insider-related ontologies, 77
- insider threat detection and prediction, 78–79
- Jaccard similarity metric, 90
- machine-learning algorithm, 79, 80
- matching process of monitored document, 95–98
- ontology-based search and information retrieval, 80–81
- ontology concept tree, 82, 96
- RNCVM, 81
- SEAM, 81
- semantic signature matching, 85–87
- steganography, 80
- structured data matching, 76
- SW algorithm, 80
- TF baseline model, 91–93
- TF-IDF baseline models, 80, 91–93
- twofold cross validation, 89
- unstructured data matching, 76

DBSCAN clustering algorithm, 31, 32
Dead nodes, 41
Delivery predictability, definition, 64
Delivery ratio (DR), 41, 89–90
Density based clustering algorithms, 31
Detect and isolate malicious host (DIMH), 195
Differential computational analysis (DCA), 154
Differential fault analysis (DFA), 154, 156–157
Dijkstra's shortest path algorithm, 63
DLP, *see* Data loss prevention
Document concept tree (DCT), 83–84
Document semantic signature (DSS), 77, 84, 96, 97
DRM schemes, 154

Dual-resolution (DR) approach, 203
Dynamic source routing (DSR) protocol, 116, 132

E

Efficient firefly routing (EFR) protocol, 134–135
- assumptions, 133
- attractiveness of, 131
- firefly-swarm next hop selection, 133, 134
- GloMoSim, 130
- light intensity, 131
- motion of a dimmer firefly, 131
- performance evaluation
  - average delay *vs.* no. of nodes, 137, 138
  - average delay *vs.* speed of nodes, 139, 140
  - end-to-end delay, 135
  - lost packet percentage *vs.* no. of nodes, 136, 138
  - lost packet percentage *vs.* speed of nodes, 138
  - packet delivery ratio *vs.* no. of nodes, 137, 138
  - packet delivery ratio *vs.* speed of nodes, 138, 139
  - simulation parameters, 136
  - throughput *vs.* no. of nodes, 137, 138
  - throughput *vs.* speed of nodes, 139
- principle of, 131
- procedural operations, 133
- retried route discovery, 135
- route discovery mechanism initiation, 134
- route reply initiation, 135
- transmission of data packets, 135
- two-step process, 132–133

EFR protocol, *see* Efficient firefly routing protocol
Elliptic curve cryptography (ECC), 49
Energy detector sensing method, 146–147
Energy-efficient versions of the PRoPHET (EPRoPHET)
- average remaining energy *vs.* number of nodes, 42–45
- delivered message *vs.* number of nodes, 42–45
- delivery predictability, 36–37
- direct probability, 37
- energy-aware mechanism, 38–39
- one-hop acknowledgment mechanism, 36
- overhead ratio *vs.* number of nodes, 44, 45
- performance matrix, 41

Index 259

simulation parameters, 39–41
transitivity formula, 37
Enron e-mail dataset, 96
EPRoPHET, *see* Energy-efficient versions of
the PRoPHET

F
False discovery rate (FDR), 79
False negative rate (FNR), 79
False positive rate (FPR), 79, 88–90
Feistel function, 158–159
"File Sink" block, 150
Financial Industry Business Ontology (FIBO),
88, 94–95
Flooding attack prevention (FAP),
195
FM radio signal, 149–150
Forbidden bands (FB), 148, 149

G
Geographical routing protocols, 233
Global positioning system (GPS), 240
GloMoSim, 130
Google assistant support, 176, 179
Greedy perimeter stateless routing (GPSR),
248
G.711 speech coders
and GSM
average voice traffic sent and received,
8–9, 11
encoder scheme, 5
end-to-end delay, 11–13
jitter, 10–12
mean opinion score, 13, 15
PDV, 12–14
quality speech, 5
setup, 6, 8
link recovery/failure, 6, 9
OSPF scenario, 4, 7, 10
RIP scenario, 7, 10
setup, 5, 7
Windows and Linux operating systems,
4

H
Hash-based message authentication code
(HMAC)
definition, 103–104
optimizations, 107–108
Hilbert–Huang transform (HHT) method, 204

I
Insider-related ontologies, 77
Insider threat detection and prediction, 78–79
IP routing protocols, 3

J
Jaccard model, 86, 98
Jittered random sampling (JRS), 144–145,
148
MED, 184–186
PD *vs.* SNR, 188, 189
PD *vs.* the smoothing factor, 188, 189
primary users, 183
random sampling mode, 186–187
ROC curve, 188
secondary users, 183
simulation diagram, 187
spectrum sensing, 184

K
Key derivation function (KDF), 101, 160
Key distribution center (KDC) scheme, 238
Key management scheme, WSN, 237–239

L
Location-based services
communication schema, 29
crypto-based privacy model, 28
features, 29
protocol schema, 30
simulation scenario
density based clustering algorithm,
31
performance metrics, 31
Weeplaces dataset, 31, 32
tracking capability, 27
Logistics study, 233

M
Malicious node breaching security, 21, 22
MANET, *see* Mobile ad hoc network
Map-based mobility model (MBM)
Dijkstra's shortest path algorithm, 63
SPMBMM, 63
Maximum eigenvalue detection (MED),
184–186, 188
Mean opinion score (MOS), 13, 15
Miller–Rabin primality test, 50
MixColumns transformation, 162

260 Index

Mobile ad hoc network (MANET)
AODV, 194
average end to end delay, 199, 200
data transmission and security, 193
packet delivery ratio, 199–200
routing, 193
RREQ, 193–195
simulation parameters, 198
throughput, 199
TSPM, 196–198
Monte Carlo method, 148–149
MQTT protocol, 168–170, 174

N
Netflix, 154
N_list format, 197
NodeMCU, 169–171, 173–175

O
Open Shortest Path First (OSPF), 3
Opportunistic network environment (ONE)
simulator, 53, 64
OppIoT, *see* Opportunistic IoT
OppNets, *see* Opportunistic networks
Opportunistic IoT (OppIoT)
access control mechanism, 49
anonymity-based authentication, 49
content-based routing, 48
elliptic curve cryptography, 49
flooding-based attacks, 49
intrusion detection approaches, 49
key-based cryptography approaches, 49
RSA-based asymmetric cryptography
decryption, 54
encryption, 51
extended Euclidean algorithm, 49
generation of keys, 51
intermediate nodes, 50
Miller–Rabin primality test, 50
private key, 50
public key, 50
RSASec protocol, 50–52
secure RSA-based routing, 50, 51
two-way authentication security
scheme, 49
simulation results
average latency *vs.* message generation
interval, 58
average latency *vs.* number of nodes, 55
average latency *vs.* TTL, 56
delivery probability *vs.* message
generation interval, 57

delivery probability *vs.* number of
nodes, 53, 54
delivery probability *vs.* TTL, 55
number of messages dropped *vs.*
message generation interval, 57
number of messages dropped *vs.*
number of nodes, 53, 54
number of messages dropped *vs.* TTL,
55, 56
ONE simulator, 53
packet delivery percentage *vs.*
percentage of malicious nodes, 53,
54
shortest path map-based mobility
model, 53
simulation parameters, 53
SMRP, 48
TSRF, 49
Opportunistic networks (OppNets)
A*OR, 62, 64
BUBBLE Rap protocol, 38
data authentication
altered/new packet detection *vs.* number
of malicious count, 24
authenticated delivery *vs.* number of
malicious count, 25
authenticated delivery *vs.* number of
nodes, 23, 24
authentication key, 20–21
CRC, 19–20
data structure and variables, 23
definition, 17
destination node, 23
fake data packets detection, 21, 22
integrity of data, 23
malicious node breaching security, 21,
22
merkle tree based technique, 19
message integrity and reliability, 18
packet alteration attack, 18
replay attack, 18
energy-aware routing scheme, 38
EPRoPHET
average remaining energy *vs.* number of
nodes, 42–45
delivered message *vs.* number of nodes,
42–45
delivery predictability, 36–37
direct probability, 37
energy-aware mechanism, 38–39
one-hop acknowledgment mechanism,
36
overhead ratio *vs.* number of nodes, 44,
45

Index   261

performance matrix, 41
simulation parameters, 39–41
transitivity formula, 37
ES&W
average remaining energy *vs.* number of
nodes, 42–45
delivered message *vs.* number of nodes,
42–45
description, 37
design steps, 39
one-hop acknowledgment mechanism,
36
overhead ratio *vs.* number of nodes, 44,
45
performance matrix, 41
simulation parameters, 39–41
spray phase, 37
vs. S&W protocol, 43–44
wait phase, 37
experiment performance metric setup, 64
fitness function, 38
MBM (*see* Map-based mobility model)
message packet forwarding mechanism, 35
next-hop selection process, 35
node's storage, 36
pairing based encryption techniques, 19
prevalence of, 35
real time traces, 63–64
RWP (*see* Random way-point model)
schematic diagram, 17, 18
secure communication, 19
store-carry-and-forward mechanism, 35
OPTICS algorithm, 31, 32
Overhead ratio (OR), 41, 64

P
Packet delay variation (PDV), 12–14
Packet delivery ratio, 199–200
Pairing based encryption techniques, 19
Pairwise key establishment scheme, 237
Particle swarm angular routing in VANETs
(PSARV) protocol
average delay *vs.* no. of nodes, 123, 125
average delay *vs.* speed of nodes, 125, 126
loss packet percentage *vs.* no. of nodes,
124, 125
loss packet percentage *vs.* speed of nodes,
125, 126
packet delivery ratio *vs.* no. of nodes, 123
route discovery initiation, 118, 119
RREQ-swarm forwarding, 119–122
simulation parameters, 122
swarm movement algorithm, 117–118

throughput *vs.* no. of nodes, 124, 125
throughput *vs.* speed of nodes, 124, 126
PBKDF2
CPU testing, 111–112
derived key, 102–103
GPU testing, 109–111
HMAC, 103–104, 107–108
iteration count, 101
optimizations, 107
password-based key derivation function,
102
PBKDF2-HMAC-SHA-1, 102
pseudorandom function, 101, 102, 106
SHA-1, 104–106, 108–109
Primary users (PU), 183
PSARV protocol, *see* Particle swarm angular
routing in VANETs protocol
Pseudorandom function (PRF), 101
Psychological profiling, 78

Q
Query quorum, 235

R
Random matrix theory (RMT), 185
Random sampling (RS) mode, 144–145,
186–187
Random way-point model (RWP)
average hop count
vs. message generation interval, 67, 68
vs. TTL, 65, 66, 70, 71
average latency
vs. message generation interval, 69
vs. TTL, 66, 67, 72
delivery probability
vs. message generation interval, 67, 68
vs. TTL, 65, 70
description, 62–63
dropped messages
vs. message generation interval, 68, 69
vs. TTL, 68, 71
overhead ratio
vs. message generation interval, 69, 70
vs. TTL, 67, 72
standard deviation, 65
Raspberry Pi 3, 170, 173, 174
Real time traces (RTT)
average hop count
vs. message generation interval, 67, 68
vs. TTL, 65, 66, 70, 71
average latency
vs. message generation interval, 69

262   Index

Real time traces (RTT) (*cont.*)
   vs. TTL, 66, 67, 72
  delivery probability
    vs. message generation interval, 67, 68
    vs. TTL, 65, 70
  dropped messages
    vs. message generation interval, 68, 69
    vs. TTL, 68, 71
  ONE simulator, 63
  overhead ratio
    vs. message generation interval, 69, 70
    vs. TTL, 67, 72
  standard deviation, 65
Real-time usage profiling, 78
Receiver operating characteristic (ROC) curve,
    147, 150, 188
Relevancy nodes-based concept vector model
    (RNCVM), 81
Route request (RREQ), 193–196
Routing Information Protocol (RIP), 3
RREQ-swarm forwarding
  Initiate the Route Discovery phase, 122
  by intermediate nodes, 119, 120
  pBestValue, 118
  PSO equations, 118
  route reply, 121–122
  route retry, 121
  schematic diagram, 117
  swarm movement algorithm, 121
RTT, *see* Real time traces
RWP, *see* Random way-point model

S
Secondary users (SU), 183
Secure criss-cross location-based service
    (SCC-LBS)
  for ad-hoc networks, 234, 235
  applications, 234
  cartesian routing, 236
  characteristics, 239–240
  client server architecture, 252–253
  DREAM technique, 234, 236
  evaluation metrics, 250
  geographical redirecting location services,
    236
  GPS, 240
  GPSR, 248
  HIGHT algorithm, 249
  Java programming, 249
  landmark system, 236
  location request, 235
  location update, 235
  location updaters selection, 241–243

  message overhead, 251, 252
  Ns-2 simulation, 248, 249
  parameters for simulation, 248, 249
  per-node database size, 252
  pseudocode, 242
  querying location updaters, 243–246
  query success rate, 251
  quorum-based approach, 235
  reactive flooding-based location service,
    234, 249
  rendezvous-based location service, 234
  security of, 246–248
  socket programming scenario, 249
  statistical performance, 250
  system architecture, 240–241
  threshold distance, 250
  vehicle P19's location updaters, 243
Secure multihop routing protocol (SMRP),
    48
Semantic signature matching,
    85–87
Semi-automated ontology management
    (SEAM), 81
Sensitive information dissemination detection
    (SIDD) system, 79
SHA-1 algorithm
  definition, 104
  graphical representation, 104
  message scheduling function, 105
  three-round optimization, 109
  word expansion phase, 108
  XORs equations, 105
  zero-based optimization, 108
ShiftRow transformation, 162
Shortest path map-based movement model
    (SPMBMM), 63
  average hop count
    vs. message generation interval, 67,
      68
    vs. TTL, 65, 66, 70, 71
  average latency
    vs. message generation interval, 69
    vs. TTL, 66, 67, 72
  delivery probability
    vs. message generation interval, 67, 68
    vs. TTL, 65, 70
  dropped messages
    vs. message generation interval, 68, 69
    vs. TTL, 68, 71
  overhead ratio
    vs. message generation interval, 69, 70
    vs. TTL, 67, 72
  standard deviation, 65
Smith–Waterman (SW) algorithm, 80

Index

Software-defined radio (SDR), 143,
    220–222
  data acquisition, 208–209
  testbed architecture, 207–208
SPACE, 155
  design, 157–158
  Feistel function, 158–159
SPARQL query, 81
Spectrum sensing (SS), 184
  analog to digital converter, 144
  coherent detection, 220
  computational resources, 221–222
  energy detector method, 146–147
  frequency domain energy detector, 144
  Monte Carlo simulation, 148–149
  neural network configuration, 227–229
  non-coherent detection, 220
  primary user's engine, 224
  random sampling mode, 144–145
  real FM radio signal, 149–150
  SDR, 143
  secondary user's engine
    cognitive cycle, 225
    continuous sensing mechanism, 226
    discrete sensing mechanism, 226
    FFT algorithm, 225
    neural network-based energy detection,
        225
    physical layer configuration, 225, 226
    state machine, flow diagram, 225, 227
  software architecture
    CRTS software, 222
    scenario controller, 222–223
    scenario test parameters, 222
    SDR-based system architecture, 223
    system node controller, 222
  SU and PU power level, 228–230
SPMBMM, *see* Shortest path map-based
    movement model
SPNbox family, 160–161
Spotify, 154
Spray-and-Wait protocol (ES&W), 43–44
  average remaining energy *vs.* number of
    nodes, 42–45
  delivered message *vs.* number of nodes,
    42–45
  description, 37
  design steps, 39
  one-hop acknowledgment mechanism, 36
  overhead ratio *vs.* number of nodes, 44, 45
  performance matrix, 41
  simulation parameters, 39–41
  spray phase, 37
  wait phase, 37

ST-based representation
  AMR, 205
  discrete model, 205–207
  dual-resolution approach
    computational time, 213, 214
    frequency resolution, 206, 207
    linear vectorial equation, 207
    normalized $L2$-norm error, 213
    objective, 205
    performance indicators, 213
    schematic diagram, 206
    signal detection and extracted contours,
        210–211
    vs. STFT, 212
    threshold, 211
    validation, 209–210
  SDR
    data acquisition, 208–209
    testbed architecture, 207–208
  time–frequency transforms,
      204–205
  wavelet transform, 205
  of wideband signal, 209–211
Steganography, 80
Structured data matching, 76
Substitution–permutation network (SPN),
    160–161
Swarm movement algorithm, 117–118

T
Term frequency and inverse document
      frequency (TF-IDF) baseline model,
      80, 91–93
Term frequency (TF) baseline model,
      91–93
TE-215 sensor, 170, 171, 173
Time–frequency transforms, 204–205
Time to Live (TTL), 196
Topology-based routing protocols, 233
Tracy-Widom distribution, 186
Trust-aware secure routing framework (TSRF),
      49
Trusted base station scheme, 238
TTP free schema, 28
Two-step protection method (TSPM),
      196–198

U
Unstructured data matching, 76
u TESLA, 238

Index

V

Vehicular ad hoc networks (VANETs)
 architecture of, 115, 129, 130
 bio-inspired routing protocols, 132
 DSR protocol, 116, 132
 EFR protocol (*see* Efficient firefly routing protocol)
 position-based routing protocols, 115–116
 PSARV protocol
  average delay *vs.* no. of nodes, 123, 125
  average delay *vs.* speed of nodes, 125, 126
  loss packet percentage *vs.* no. of nodes, 124, 125
  loss packet percentage *vs.* speed of nodes, 125, 126
  packet delivery ratio *vs.* no. of nodes, 123
  route discovery initiation, 118, 119
  RREQ-swarm forwarding, 119–122
  simulation parameters, 122
  swarm movement algorithm, 117–118
  throughput *vs.* no. of nodes, 124, 125
  throughput *vs.* speed of nodes, 124, 126
 swarm intelligent-based routing protocols, 116
Voice encoding codec
 features, 2
 G.711 speech coders (*see* G.711 speech coders)
 types, 2, 3
VoIP
 application configuration, 5
 audio and speech compression, 4
 call quality, 3
 description, 1
 link recovery configuration, 6, 9
 network topology, 4, 5

OPNET modeler, 4
 profile configuration, 5, 6
 QoS, 1
 route reconvergence, 4
 routing protocols, 2–3
 signal quality, 4
 voice encoding codec
  features, 2
  G.711 speech coders (*see* G.711 speech coders)
  types, 2, 3
VT-CORNET, 221, 222

W

Weeplaces dataset, 31, 32
White-box cryptography
 AES implementation, 154–156
 AES-NI instructions, 163
 ASASA, 155
 BitShift transformation, 162–164
 DCA attack, 154, 156–157
 definition, 154
 DES and AES, 154, 155
 DFA attack, 156–157
 issues, 162
 performance evaluation, 164
 SPACE, 155
  design, 157–158
  Feistel function, 158–159
 SPNbox family, 160–161
Wigner-Ville distribution (WVD), 204
WordNet, 81
"WX GUI FFT Sink" block, 150

Z

Zero difference enumeration (ZDE), 154

CPSIA information can be obtained
at www.ICGtesting.com
Printed in the USA
LVHW011627310319
612460LV00003B/202/P